D1759863

**Essays in
Plant Taxonomy**

Essays in Plant Taxonomy

Edited by

H. E. STREET

Botanical Laboratories, University of Leicester

1978

ACADEMIC PRESS
London   New York   San Francisco
A Subsidiary of Harcourt Brace Jovanovich, Publishers

ACADEMIC PRESS INC. (LONDON) LTD.
24/28 Oval Road,
London NW1

United States Edition published by
ACADEMIC PRESS INC.
111 Fifth Avenue
New York, New York 10003

Library of Congress Catalog Card Number: 77 15335
ISBN: 0 12 673360 0

PRINTED IN GREAT BRITAIN BY
WILLMER BROTHERS LIMITED BIRKENHEAD

Contributors

A. R. CLAPHAM, *The Parrock, Arkholme, Carnforth, Lancashire, England*
A. O. CHATER, *Department of Botany, British Museum (Natural History), Cromwell Road, London SW7, England*
C. D. K. COOK, *Institut für Systematische Botanik, Universität Zurich, 8039 Zurich, Switzerland*
J. G. HAWKES, *Department of Botany, University of Birmingham, P.O. Box 363, Birmingham, England*
D. L. HAWKSWORTH, *Commonwealth Mycological Institute, Ferry Lane, Kew, Richmond, Surrey, England*
V. H. HEYWOOD, *Department of Botany, University of Reading, Whiteknights, Reading, Berkshire, England*
D. M. MOORE, *Department of Botany, University of Reading, Whiteknights, Reading, Berkshire, England*
P. F. PARKER, *Botanical Laboratories, University of Leicester, Leicester, England*
P. W. RICHARDS, *14 Wootton Way, Cambridge, England*
I. B. K. RICHARDSON, *Royal Botanic Gardens, Kew, Richmond, Surrey, England*
P. M. SMITH, *Department of Botany, University of Edinburgh, King's Buildings, Mayfield Road, Edinburgh, Scotland*
P. H. A. SNEATH, *Department of Microbiology, University of Leicester, Leicester, England*
C. A. STACE, *Botanical Laboratories, University of Leicester, Leicester, England*
D. H. VALENTINE, *Department of Botany, University of Manchester, Oxford Road, Manchester, England*
S. M. WALTERS, *University Botanic Garden, Brookside, Cambridge, England*
P. F. YEO, *University Botanic Garden, Brookside, Cambridge, England*

v

Publisher's Note

This work was conceived and promoted by Professor H. E. Street in the winter of 1976–1977. He completed reading and checking the page-proofs immediately before his tragic and sudden death on 4th December 1977, and this book represents one of the last of the hundreds of scientific publications with which he was involved.

Preface

This volume seeks to present a collection of essays which together survey some of the major problems and fields of active research in plant taxonomy. Thereby it is hoped that students will have access to critical reviews which will supplement and extend their formal teaching in this subject area and that a wider audience of biologists will find here a readable and up-to-date assessment of this foundation area of plant science.

I am deeply grateful to my authors particularly since they cheerfully conformed to a very tight schedule to ensure publication of the volume at the appropriate time. Each is a personal friend and some are former students of Tom Tutin; the association of this volume with his 70th Birthday prompted them in a number of cases to set aside other obligations to ensure the receipt of their manuscripts on time.

I took on the Editorship because I was convinced of the potential value of such a volume and because of the opportunity it offered to pay tribute to my colleague and former holder of the Chair of Botany here in Leicester. I need hardly say that I am singularly unfitted to the task of editing, in any academic sense, a work on Plant Taxonomy. My task was, however, made easy by the care exercised by my authors in preparing their manuscripts and by the expert assistance, so freely given, which I received from my colleague Dr Clive Stace.

June, 1977   H. E. STREET

Thomas Gaskell Tutin

Thomas Gaskell Tutin was the only son of Frank and Jane Tutin, but he has one sister, Elizabeth, 4 years younger than himself. His parents were second cousins, his father having married Jane Arden of Knutsford, a relative of his own mother, née Elizabeth Arden. Through Thomas Gaskell, clockmaker in Knutsford in the eighteenth century, whose daughter married an Arden, there were family connections with Mrs Gaskell, authoress of *Cranford*.

Tom's grandfather Tutin led a highly eventful life before becoming a successful builder and contractor in Nottingham. After he had been bankrupted through an unfortunate law-suit, his son, Tom's father, felt free to discontinue his training as an architect and followed his own inclination by making chemistry his profession. He left Nottingham for London and became a biochemist with the Wellcome Foundation and later at the Lister Institute. After their marriage in 1904 he and his wife went to live in Kew, where Tom was born on 21st April, 1908. Tom's sister remembers that on Sunday mornings she was often taken in a push-chair by her father and brother when they walked along the river between Kew and Richmond looking for plants and aquatic animals. Tom's father was deeply interested in all branches of natural history: as a boy he had grown ferns in the greenhouse of the large garden outside Nottingham where his father raised prize vegetables, fruit and roses. He was a keen and knowledgeable field botanist and had a sizeable private herbarium: his son inherited a well-used *Bentham and Hooker*. With this paternal encouragement Tom developed an early interest in wild plants and also in moths and butterflies. At this time, too, there were many visits to Kew Gardens. Elizabeth Tutin tells of a day when two of the gardeners, who were clearing a bed and had thrown out a heap of plant material, asked the young Tom what he thought it was. To their astonishment he identified it as rhizomes of

Anemone apennina, and was given some for his garden. At about this time he discovered *Drosera rotundifolia* on Sheen Common and reported his find to the appropriate authorities.

In March 1920, the family moved to Flax Bourton near Bristol. Tom's father had been told that he must leave London for the sake of his health, and he obtained a post as biochemist at Long Ashton Research Station. Tom went to Cotham Secondary School, chosen for its good science teaching. His enthusiasm for natural history was undiminished and there were many family excursions into the Bristol countryside.

In 1927 Tom went to Cambridge as a Scholar of Downing and began reading for Part I of the Natural Sciences Tripos. By the end of his first year he wrote that it had become clear to him "that whatever I ultimately did, it would not be botany". At the beginning of the Long Vacation of 1928, immediately following that resolve, he met Humphrey Gilbert-Carter, the idiosyncratic and lovable Director of the Cambridge Botanic Garden, and by the end of the 6-week term he "had learnt that Botany is a special way of life linking the Arts and the Sciences". He became a devoted disciple of Gilbert-Carter, spending much time on the famous botanical excursions and at Cory Lodge. He also reversed his earlier decision against a career in botany and in due course took it as his subject in Part II of the Tripos.

As an undergraduate Tom became much involved with the Cambridge Natural History Society, as did his contemporaries P. W. Richards and E. F. Warburg, of whom he saw a great deal from the beginning of his second year onwards. All three also became members of the Natural Science Club, contributing no fewer than 19 papers between them. By this time Tom was an outstandingly able field botanist and was actively developing his interest in sedges and especially grasses. In 1929 their friend A. P. F. Michelmore suggested an expedition to the Azores in which both Tom and Warburg took part during the Long Vacation between their second and third years.

Tom took Part II in 1930 and stayed on in Cambridge to work on some fossil plants brought back from Greenland by Professor A. C. Seward. In the spring of 1931 he went with P. W. Richards and Dr W. Balfour Gourlay on a plant-collecting trip to southern Spain and Spanish Morocco, and in 1933 joined an expedition to British Guiana. On his return he went to the Marine Biological Station at

Plymouth to investigate the wasting disease of *Zostera* that was causing concern on both sides of the Atlantic because of the importance of eel-grass as food for water-fowl. In the course of his work he added a new species of *Zostera* to the British flora and published three papers on the genus, followed later (1942) by an account for the *Biological Flora of the British Isles*. He left Plymouth in 1937 to take part in the expedition to Lake Titicaca led by H. C. Gilson and supported by the Percy Sladen Trust. He contributed the Expedition reports on the macrophytic vegetation and the algae of the Lake. His observations and reflections led in 1941 to a paper in the *Journal of Ecology* entitled *The Hydrosere and Current Concepts of the Climax*. He pointed out that a large deep lake, with sufficient wave action for incoming silt to be well distributed over the deeper parts of the floor but with no excessive turbulence, will have areas of floor which are adequately illuminated and sufficiently stable for submerged rooted vegetation to persist unchanged for indefinitely long periods, so that it may fairly be termed "climax". It is only in sheltered bays that true hydroseres can give rise to the familiar sequence through floating-leaved, reed-swamp and fen-wood communities to a *terrestrial* climax vegetation. But the earliest stages of such seres are typically separated from the *aquatic* climax by a bare zone maintained by wave action and have no successional connection with it. This was a timely contribution to ecological thinking when the credibility of Clements' oversimplified successional theory was being questioned.

On his return from Lake Titicaca, Tom first held a part-time post at King's College, London, and then, in 1939, was appointed Assistant Lecturer at Manchester. He now spent a good deal of time at Wray Castle, by Windermere, developing the interest in algal plankton that began during the Titicaca excursion. It was there that he met Winifred Pennington, his future wife, and they were married in 1942. In that year he joined the Geographical Section of the Naval Intelligence Division of the Admiralty, which was engaged in the war-time task of preparing a new series of geographical handbooks. Then in 1944 he made his final move on being appointed Lecturer-in-Charge of the Botany Department at Leicester University College, becoming the first Professor of Botany when the College received its Charter as an independent University in 1947. Twenty years later, when the School of Biological Sciences was established,

Tom became the first occupant of a newly created Chair of Taxonomy, from which he retired in 1973. He is now Emeritus Professor and a University Research Fellow.

It was on a winter afternoon just after the war that Tom went for a walk with Humphrey Gilbert-Carter, and when they reached Grantchester they decided to call on A. G. Tansley. Over tea Tansley suggested that Tom should write a new British Flora: he had long felt that the lack of a modern working Flora was seriously hampering both the teaching and the learning of field botany and ecology. Tom was immediately interested in the idea, plans were discussed on the spot, and the *Flora of the British Isles* was completed within 4 years of that tea-time conversation, though it did not appear in print until 1952. The leading part Tom Tutin played was made clear at the time: "Two of us wish at this point to acknowledge the special contribution of T. G. Tutin, who besides writing a substantial part of this flora undertook in addition the arduous task of acting as general editor. It was he who collected and collated the various sections as they were completed, who strove to secure uniformity of treatment, who wrestled with text-figures, glossary and index, and who urged us on when we flagged".

Apart from assuming the usual responsibilities of a university department, Tom's special interests led him to develop a Botanical Garden from 1945 onwards and to start on a Herbarium in 1946. A quarter of a century later he was able to report, with justifiable pride, that the Herbarium housed nearly 80,000 specimens of European flowering plants, gymnosperms and ferns with a smaller number of bryophytes, lichens and algae, as well as Dr D. M. Moore's valuable collection of plants from the Falkland Islands and the southern part of South America. Visitors from 17 countries had come to consult the Leicester Herbarium up to 1971, an impressive indication of its international standing.

At the 8th International Botanical Congress, held in Paris in 1954, there was a symposium on *Progress of Work on the European Flora*, organized by Professor D. H. Valentine. In his introductory talk he referred to the several new national Floras already published or in course of preparation and added "a new European flora, though an immense undertaking, must be regarded as one of the aims of the future". Tom Tutin followed up this proposal with characteristic energy and enthusiasm, discussing it with botanists from many countries and inducing N. A. Burges, V. H. Heywood, D. H. Valentine, S. M. Walters and D. A. Webb to join him on what became an informal editorial committee for *Flora Europaea*, the first full meeting being held in Leicester early in 1956. It seems clear that without Tom Tutin *Flora Europaea* would never have got off the ground. He provided the necessary stimulus and driving force, and there was never any doubt that he was the only possible chairman of that editorial committee which, with the early addition of Dr D. M. Moore, eventually became the formal executive body in charge of the project. It was an obvious advantage that it should be small and readily convened, but steps had to be taken to ensure close and continuous contact with collaborators in other countries. Now, after more than 20 years of determined hard work, the fifth and final volume is due to appear in the same year as this book. Although the aim was to produce a complete Flora in the shortest possible time and not to delay publication to allow the solution of all the problems that might arise, *Flora Europaea* is nevertheless much more than a compilation based only on previously existing knowledge. The Czech botanist J. Holub, writing in 1966, recognized the first volume as incorporating results of new studies which have greatly deepened our understanding of a number of taxonomic groups. There can be no doubt that *Flora Europaea* is very much to the credit of all concerned and especially of Tom Tutin, who has probably contributed more, as editor and as author, than anyone else and whose patience, firmness and good humour have made him an admirable chairman through-out. And he has continued that work with unflagging zeal despite his retirement over 4 years ago. At the Anniversary Meeting of the Linnean Society of London on the 24th May, 1977, Tom's many distinguished contributions to plant taxonomy were marked by the presentation of the Society's Gold Medal, which is awarded to leading biologists "as an expression of the Society's estimate of their services to science".

This short introduction must necessarily leave much unwritten. Many will know the comfortable house on the outskirts of Leicester where the Tutins brought up a son and three daughters and entertained their friends, and where Tom spent much of his weekend in gardening until, sadly, this became a forbidden occupation. It seems that he much enjoyed cricket at school and at Flax Bourton, but his chief outdoor activity has surely been walking in search of plants, and he must have covered thousands of miles in doing so. He has, perhaps, preferred botanizing on more or less level ground to climbing mountains, but this view may be unduly coloured by memories of a hot and tiring afternoon on the Sow of Atholl. Tom had looked at the long slope and decided against ascending it, but he gave the two other members of the party a compass-bearing that should have taken them straight to the patch of *Phyllodoce caerulea*, but did not in fact do so.

When darkness, inclement weather, stress of circumstances or mere fatigue have prevented him from walking after plants, Tom has always been happy to talk about them over a glass of beer. His more complete leisure has been spent in reading or in indulging that love of music he derived from his mother, perhaps listening to a Mozart opera, perhaps playing his flute.

This book has been prepared in celebration of his 70th birthday by a number of Tom's many friends who seek in this way to show their admiration for his achievements as a botanist and their affection for him as a person. They wish him very many more years of delight in plants and enjoyment of life in general.

I wish finally to express my deep gratitude to all those who have helped me with information for this biographical note.

July, 1977   A. R. CLAPHAM

Publications of T. G. Tutin

1932   (with E. F. Warburg). Notes on the flora of the Azores. *J. Bot., Lond.* **70**, 7–13, 38–46.

A cretaceous Gleicheniaceous fern from Greenland. *Ann. Bot.* **46**, 503–508.

(with T. T. Macan). A note on rot-holes in horse chestnut trees. *Parasitology* **24**, 283.

1933   On *Ilex perado* Ait. and *Notelea excelsa*. *J. Bot., Lond.* **71**, 99–101.

(with J. S. L. Gilmour). "A list of the Important Collections in the University Herbarium, Cambridge". Cambridge University Press, Cambridge.

1934   New species from British Guiana, Cambridge University Expedition, 1933. *J. Bot., Lond.* **72**, 306–314.

The fungus on *Zostera marina*. *Nature, Lond.* **134**, 573.

1935   Species novae vel minus cognitae a T. G. Tutin descriptae. *Hooker's Icones Plantarum*, Ser. 5, **3**, 3273–3275.

Bromus lepidus, Holmb. *J. Bot., Lond.* **73**, 235.

1936   A revision of the genus *Pariana* (Gramineae). *J. Linn. Soc., Bot.* **50**, 337–362.

New species of *Zostera* from Britain *J. Bot., Lond.* **74**, 227–230.

Bartsia viscosa in Norfolk. *J. Bot., Lond.* **74**, 297.

1937   *Zostera Hornemanniana* Tutin in Scandinavia. *Svensk Bot. Tidskr.* **31**, 215–216.

1938   The autecology of *Zostera marina* in relation to its wasting disease. *New Phytol.* **37**, 50–71. *Nature, Lond.* **141**, 1147.

1939   *Zostera Hornemanniana* from Caernarvonshire. *J. Bot., Lond.* **77**, 96.

1940   Reports of the Percy Sladen expedition to Lake Titicaca. X. The macrophytic vegetation of the Lake. XI. The algae. *Trans. Linn. Soc. Lond.*, Ser. 3, **1**, 161–202.

New species from British Guiana. *J. Bot., Lond.* **78**, 249–257.

1941   The hydrosere and current concepts of the climax. *J. Ecol.* **29**, 268–279.

1942 *Zostera* L. In Biological Flora of the British Isles. *J. Ecol.* **30**, 217–226.

1950 A note on species pairs in the Gramineae. *Watsonia* **1**, 224–227. *Milium scabrum* Merlet. *Watsonia* **1**, 345–348.

1952 Origin of *Poa annua*. *Nature, Lond.* **169**, 160. Note on the nomenclature of *Roegneria doniana* (F. B. White) Meld. *Watsonia* **2**, 186–187. "The Botanical Revolution. An Inaugural Lecture". University College, Leicester. (with A. R. Clapham and E. F. Warburg). "Flora of the British Isles". Cambridge University Press, Cambridge.

1953 The vegetation of the Azores. *J. Ecol.* **41**, 53–61. Some general aspects of the *Zostera* problem. *Proc. VIIth Intern. Bot. Congress, Stockholm* (1950), 733–735. Natural factors contributing to a change in our flora. *In* "The Changing Flora of Britain" (J. E. Lousley ed.), pp. 19–25. Botanical Society of the British Isles.

1954 The relationship of *Poa annua* L. *In* "Rapports et Communications VIIIth Congrès International de Botanique", Sections 9 and 10, p. 88. Paris. The need for international co-operation in experimental studies on the European Flora. *In* "Rapports et Communications VIIIth Congrès International de Botanique", Sections 2, 4, 5 and 6, p. 102. Paris.

1955 Species problems in plants with reduced floral structure. *In* "Species Studies in the British Flora" (J. E. Lousley, ed) pp. 21–26. Botanical Soc. British Isles.

1956 The genus *Symphytum* in Britain. *Watsonia* **3**, 280–281. Generic criteria in flowering plants. *Watsonia* **3**, 317–323.

1957 "Flora of the British Isles—Illustrations Part I". Drawings by S. J. Roles. Cambridge University Press, Cambridge. A contribution to the experimental taxonomy of *Poa annua* L. *Watsonia* **4**, 1–10. *Allium ursinum* L. *In* Biological Flora of the British Isles. *J. Ecol.* **45**, 1003–1010.

1958 Classification of the legumes. *In* "Nutrition of the Legumes" (E. G. Hallsworth, ed.) pp. 3–14. Butterworths, London.

1959 (with A. R. Clapham and E. F. Warburg). "Excursion Flora of the British Isles". Cambridge University Press, Cambridge. (with P. W. Ball). Notes on the annual species of *Salicornia* in Britain. *Watsonia* **4**, 193–205.

1960 The centenary of the Origin of Species. *Nature, Lond.* **175**, 216–217.

Introduction: on genera and generic criteria in problems of taxonomy and distribution in the European flora. *Feddes Repert.* **63**, 113–116.
"Flora of the British Isles—Illustrations Part II". Drawings by S. J. Roles. Cambridge University Press, Cambridge.

1962 (with A. R. Clapham and E. F. Warburg). "Flora of the British Isles", 2nd ed. Cambridge University Press, Cambridge.

1963 New taxa and new names in European *Dianthus*. *In* Notulae systematicae ad Floram Europaeam spectantes 2. *Feddes Repert.* **68**, 189–193.
Scale effects and other subjective influences in taxonomy. Bot. *Notiser* **116**, 122–126.
"Flora of the British Isles—Illustrations Part III". Drawings by S. J. Roles. Cambridge University Press, Cambridge.

1964 New taxa and new names in European Ranunculaceae. *In* Notulae systematicae ad Floram Europaeam spectantes 3. *Feddes Repert.* **69**, 53–55.
Various genera including *Equisetum, Ulmus, Dianthus, Anemone, Ranunculus, Aquilegia* and *Thalictrum*. *In* "Flora Europaea" (T. G. Tutin *et al.*, eds.), Vol. 1. Cambridge University Press, Cambridge.

1965 "Flora of the British Isles—Illustrations Part IV". Drawings by S. J. Roles. Cambridge University Press, Cambridge.
Recent advances in European taxonomy—the general picture. *Rev. roum. Biol.* **10**, 11–17.
Notulae. *In* Notulae systematicae ad Floram Europaeam spectantes 5. *Feddes Repert.* **70**, 4–5.

1967 Short Notulae, in Notulae systematicae ad Floram Europaeam spectantes 6. *Feddes Repert.* **74**, 25, 26, 31–34.

1968 Notes on the genus *Athamanta* (Umbelliferae) in Europe. *In* Notulae systematicae ad Floram Europaeam spectantes 7. *Feddes Repert.* **79**, 18–20.
Short Notulae. *In* Notulae systematicae ad Floram Europaeam spectantes 7. *Feddes Repert.* **79**, 53, 54–55, 56, 61–62.
(with A. R. Clapham). "Excursion Flora of the British Isles", 2nd ed. Cambridge University Press, Cambridge.
(with A. C. Jermy). "British Sedges". Botanical Soc. British Isles.
Some European species of *Medicago*. *Collectanea Botanica,* Barcelona **7**, 1163–1166.
Various genera including *Medicago, Euphorbia, Pimpinella, Bupleurum, Peucedanum* and *Laserpitium*. *In* "Flora Europaea" (T. G. Tutin *et al.*, eds), Vol. 2. Cambridge University Press, Cambridge.

1970   *Gentiana* sect. *Megalanthe* Gaudin. *Fragmenta Floristica et Geobotanica* **16**, 85–90.

1971   Short Notes. *In* Notulae systematicae ad Floram Europaeam spectantes 11. *Bot. J. Linn. Soc.* **64**, 378.

1972   Short Notes. *In* Notulae systematicae ad Floram Europaeam spectantes 12. *Bot. J. Linn. Soc.* **65**, 259–260, 262.
       Various genera including *Gentiana*, *Gentianella*, *Teucrium* and *Globularia*. *In* "Flora Europaea" (T. G. Tutin *et al.*, eds), Vol. 3. Cambridge University Press, Cambridge.

1973   Weeds of a Leicester garden. *Watsonia* **9**, 263–267.
       Notes on *Jasione* in Europe. *In* Notulae systematicae ad Floram Europaeam spectantes 14. *Bot. J. Linn. Soc.* **67**, 276–279.
       Short Notes. *In* Notulae systematicae ad Floram Europaeam spectantes 14. *Bot. J. Linn. Soc.* **67**, 280, 282–283.

1974   (with A. O. Chater). *Asperula occidentalis* in the British Isles. *Watsonia* **10**, 170–171.

1975   *Molinia* in S.W. Spain. *Lagascalia* **5**, 73–75.
       Short Notes. *In* Notulae systematicae ad Floram Europaeam spectantes 16. *Bot. J. Linn. Soc.* **70**, 17–18.
       Short Notes. *In* Notulae systematicae ad Floram Europaeam spectantes 19. *Bot. J. Linn. Soc.* **71**, 274.

1976   Various genera including *Jasione*, *Bidens* and *Artemisia*. *In* "Flora Europaea" (T. G. Tutin *et al.*, eds), Vol. 4. Cambridge University Press, Cambridge.

The Editor apologizes for any omissions from this list which was compiled without consultation with the author.

Contents

Chapter 1. **Ecological Criteria in Plant Taxonomy**
         D. H. VALENTINE

Chapter 2. **Chemical Evidence in Plant Taxonomy**
         P. M. SMITH

Chapter 13. **British Endemics**
S. M. WALTERS

Chapter 14. **European Floristics: Past, Present and Future**
V. H. HEYWOOD

*Dedicated to
Emeritus Professor Thomas G. Tutin
on the occasion of his
70th Birthday*

Chapter 1

Ecological Criteria in Plant Taxonomy

D. H. VALENTINE

INTRODUCTION

It is probably true to say that ecological criteria are of comparatively little direct importance in taxonomy. The prime criterion, in classical taxonomy at least, has been and will doubtless remain morphological, though ecological criteria at the infraspecific level cannot be neglected. On the other hand, from an evolutionary point of view, habitat and environmental factors must always play a vital part; and evolution is the fundamental process which shapes taxonomic differentiation. Hence, in this chapter, it will be necessary to look at the evolutionary role of ecological factors, as well as their more direct taxonomic significance. We shall begin with an examination of phenotypic plasticity and tolerance and lead on from this to concepts of ecotype and ecocline in population differentiation, and the extent to which these can be taxonomically useful. Subsequently, we shall look at the significance of ecological isolation, and note the importance of the events which can occur when closely related taxa meet and hybridize at an ecotone.

PLASTICITY AND TOLERANCE

The tolerance of a plant population is measured by its ability to survive and reproduce when exposed to a range of environmental conditions; the wider the range, the greater the tolerance. The plasticity of the population is measured by the extent to which the appearance and characters of the plants vary in moving from one set of environmental conditions to another. Towards the end of the nineteenth century, when Darwin's ideas on variation and evolution were taking root, a number of botanists were led to experiment, on a field scale, on the effects of the environment on plant populations. One well known series of experiments was carried out by the Austrian botanist, Kerner (1895).

Kerner took seed samples of lowland species, both annual and perennial, and grew them in two experimental gardens. One of these was in Vienna, in the lowlands (150m), the other in the Tyrol, in a montane region (2900m) with much severer conditions and a shorter growing season. Many species were tried. It was found that many of the annuals did not survive beyond the first season in the montane garden, though a few, such as *Senecio vulgaris* L., were tolerant and produced ripe seed. Some did not flower but persisted in the vegetative state for several seasons, i.e. reacted plastically and became perennials (e.g. *Poa annua* L.). Of the perennials, only 32 out of the 300 tested flowered at high altitude. All showed (as had the annuals) marked phenotypic modifications, such as shorter stems, smaller leaves, fewer and smaller flowers and more anthocyanin (e.g. *Geum urbanum* L.). Kerner gathered seeds from some of these montane transplants, sowed them in Vienna, and showed that the plants reverted to their normal, lowland condition. A famous series of experiments on similar lines was conducted by Clausen *et al.* (1940) in California.

Kerner's experiments made it quite clear that plasticity and tolerance are widespread in flowering plants; and indeed it had long been recognized that many species were able to grow and thrive in a variety of plant communities. Certain simple generalizations could also be made about the kinds of plastic modifications that are found, e.g. the dwarfing effect of very dry conditions, and the production of succulent leaves in maritime habitats affected by salt.

ECOTYPIC AND ECOCLINAL VARIATION

The question arises, are such modifications purely plastic, or do they also have a genotypic component? The question was posed and answered by Turesson (1922b) who systematically investigated ecological variation in a large number of tolerant species of north-west Europe. Turesson had the advantage over Kerner that, in the interim, modern genetics had been founded, and the concepts of genotype and phenotype formulated by Johannsen.

Turesson's method was to grow the plants, either as transplants or from seed, in an experimental garden. Plants from different habitats could thus be grown side by side under comparable conditions. The plastic stimulus would be identical and it could be inferred that residual differences were genotypic. It should be noted that genotypic differences can only be proved from seed samples. Transplants themselves may be very slow to change in cultivation, and it is not safe to assume that plastic differences will disappear quickly. In general, Turesson found that differences between plants from different habitats were partly plastic but often also had a genotypic basis; thus some but not all of the differences disappeared in the experimental garden.

An example which has been studied in Britain by A. C. Tallantire (unpublished) is *Geum rivale* L., which occurs in a series of habitats from lowland woodland on rather wet base-rich soil to upland pastures on base-rich soils in regions of high rainfall, where it is often grazed. The extreme habitats are very different, and so are the plants. The lowland plants are about 40 cm tall, with a rather lax habit and large leaves, and they show little change when cultivated in a lowland garden. Montane plants from Upper Teesdale (500m) or the Scottish mountains (1100m) are dwarf (5 cm), compact, with small leaves and short stems, and are more hairy; and though they increase somewhat in size in the lowland garden, they retain their distinctive characters. Breeding experiments have shown that the differences are maintained in the next generation, and hybridization experiments have shown that the two kinds of plant are perfectly interfertile.

Examples of this kind are not uncommon, not only in the British Isles but in other areas which have been investigated; and in many cases, as in this one, the differences observed may be assumed to be adaptive. Thus the low-growing habit and small size of the montane plants of *G. rivale* may be regarded as reactions to wind exposure and also perhaps to grazing. The reactions are partly plastic, but some of the characters have become fixed genetically. It should be noted that such differences do not always have a genotypic component. Thus Salisbury (1940) showed that the dwarf var. *pygmaea* Lange of

Plantago coronopus L., from brackish dune-slacks, when cultivated in the experimental garden, reverted to the normal variety.

A second example, from a different kind of habitat, is that of *Geranium robertianum* L. investigated by Baker (1956). This is a widespread species in Britain, usually to be found in shady habitats and on fairly base-rich soils. It is common in the lowlands and ascends to 700m; it is also phenotypically very variable, e.g. in hairiness and breadth of leaf segments, but most of this variation is not genotypic. On shingly or stony beaches by the sea, however, a distinct population is sometimes found which differs from the inland populations in having a prostrate habit, smaller leaves and flowers, and in being later flowering. These characters are maintained in cultivation and breed true. When the maritime and inland plants are crossed, fertile offspring are obtained, and segregation is free in the F_2.

In order to avoid confusion, and to emphasize his ecological approach to the problems of variation, Turesson (1922a) devised a special terminology to present his results. The species, as recognized by classical taxonomy, the so-called "Linnean" species, he renamed the ecospecies; and he coined a new term, the ecotype, which he defined as the product of the genotypical reaction of the ecospecies to a particular, defined set of environmental conditions. Later Turesson modified his concept of ecospecies, and redefined it in terms of ability of the component parts to exchange genes freely; when this was brought into the ecotype concept, it was sometimes held to eliminate polyploid chromosome races. More recently this distinction has fallen into disuse, and both homoploid and polyploid ecotypes may be recognized.

Subsequently, more detailed study of variation over a range of habitats showed that distinct ecotypes were often difficult to characterize and that instead there was a gradient of variation correlated with a gradient of habitat conditions. Such a gradient became known as a cline (Huxley, 1938) and when linked with ecological conditions, as an ecocline. In such cases, methods of investigation and analysis have to be carefully controlled, and this is well illustrated in the experiments of Gregor and his colleagues (Gregor, 1938, 1946) on the widespread maritime species *Plantago maritima* L. This work was done mainly on Scottish populations, the plants being diploid and outbreeding. The habitats investigated were a series from muddy salt-marsh to cliff-top. Seed samples were taken from the populations and the offspring grown in an experimental garden. They were scored for a series of characters, especially plant height and habit grade (decumbent to erect), and it was shown that there was a well marked gradient in these characters as the habitat changed from one end of the series to the other. Populations at points on the cline were significantly different, although there was overlap between individuals from different populations.

It is interesting that, although the plants are wind-pollinated outbreeders, and the populations concerned were in close proximity, selection was strong enough to produce the significant genotypic differences which were observed. Gregor's first set of results, obtained on the Firth of Forth on the east coast, were confirmed by a second series taken at Lewis on the west coast, showing a similar, though not identical, ecocline.

The kinds of habitat which offer opportunities for ecotype development were first analysed by Turesson, and he was able to show that certain kinds of habitat are important in this respect, e.g. the montane, the maritime and the woodland habitat. He was thus able to suggest that the ecotypes of various species developing in such places should bear a common name, such as *oecotypus maritimus*, or *oecotypus salinus*. Where there is a combination of special conditions (including, of course, climatic conditions, which can never be entirely disentangled from edaphic), formation of ecotypes may be very noticeable. One striking example, which has not yet been systematically studied, is that of the limestone cliffs on the south shore of the Gower peninsula, near Swansea. Here are found a maritime ecotype of *Centaurea scabiosa* L. (sometimes known as var. *succisaefolia* E. S. Marshall), a maritime ecotype of *Dactylis glomerata* L. (to which the name var. *collina* Schlecht. has been applied), and what appear to be ecotypes of *Prunus spinosa* L. and *Ligustrum vulgare* L., characterized by their dwarf, decumbent habit. The genotypic basis of the first two examples has been established, that of the other two is as yet only inferred.

We should mention here an interesting group of edaphic ecotypes which have been much studied in recent years. Here the populations concerned are differentiated by their ability to grow and survive in soils with a surplus or a deficiency of a particular element. The most striking examples are found on serpentine soils, which often support an impoverished flora of specialized plants which can tolerate high concentrations of magnesium, chromium and nickel, and low concentrations of calcium and phosphorus (Kruckeberg, 1954; Proctor, 1971). Other examples are found in several common grasses, such as *Festuca ovina* L. and *Agrostis tenuis* Sibth., certain stocks of which have become adapted to living on soils rich in lead and copper (Wilkins, 1960; McNeilly, 1968). Thus lead concentrations of the order of 1000 p.p.m., lethal to normal plants, can be tolerated. The plants can be tested experimentally by growing rooted seedlings or cuttings in water culture. The tolerance is genotypic, and is not, at least in *F. ovina*, correlated with any morphological character, nor with chromosome number. Very often, too, the variation is ecoclinal.

We should also draw attention to the question of geographical variation. In the example at the beginning of this chapter, the work of Kerner, who compared the environment of the mountain and the plain, was used to illustrate the concepts of tolerance and plasticity. This lowland–montane difference is at least in part geographical; and it has often been used in field experimentation on variation. The effects produced by eco-geographical variation of this kind are analogous to those produced by strictly ecological variation as between, say, one lowland community and another, and they may be very considerable. Thus Clausen *et al.* (1940) were able to demonstrate a large number of morphological and physiological differences between lowland and alpine races of *Potentilla glandulosa* Lindl., and to show that they were controlled by about 100 genes. There was no difficulty in describing these races in taxonomic terms; indeed they had been recognized and named as taxa, either at the subspecific or the specific level, before the experimental work was done.

Turesson, too, did many experiments in which he compared ecotypes of arctic and temperate regions, or of north and south Europe, or of oceanic western Europe with the continental east, and he was able to establish "ecotypic" differences, just as he had for the different habitats in the early experiments. The "allopatric ecotypes" seemed to be formed by a reaction to climatic differences, just as the "sympatric ecotypes" were formed by a reaction to more strictly ecological factors, such as differences in soil, kind of community, exposure to wind etc. Both kinds, too, were liable to show clinal variation.

TAXONOMIC TREATMENT OF INFRASPECIFIC VARIATION

It must be agreed that there are some sorts of populations, particularly those where hybridization is frequent, in which any attempt to apply an orthodox infraspecific classification would be doomed to failure, or at least do more harm than good. On the other hand, in less complex situations, this need not be the case, and there may be advantages in an orthodox taxonomic classification in terms of subspecies, variety and form. This may present a picture not too far removed from the field situation, which will provide the general botanist with terms of reference, and enable information about variation to be communicated from one worker to another not only in space, but also in time. With this in mind, we shall say something first about the plant populations and how they can best be named, and shall follow with some remarks about habitat.

Plant populations

From the point of view of discussion of population variation in general terms, the deme terminology, proposed by Gilmour and Gregor (1939) has many advantages, and its use has been well illustrated in the book of Briggs and Walters (1969). It is often very convenient to refer to a freely outbreeding population as a gamodeme, or a population in a particular locality as a topodeme; and one can particularize to a certain extent, and speak, for instance, of a sand-dune ecodeme, or even, say, an Ainsdale sand-dune ecodeme. But such a treatment has neither the precision nor the authority of a formal taxonomic description, when it is a question of dealing with populations which need to be recognized and identified by all and sundry. For such a purpose, we need a stable name in a recognized hierarchy; and, with all its imperfections, the term variety seems the best to use for Turesson's ecotype. This makes no allowance for the existence of clines, and it may lead to trouble when ecotypes are spread over a large geographical area, and questions of polytopic origin have to be considered. Nevertheless, the naming of ecotypes as varieties would provide students of ecology with a useful body of information; and so long as ecotypes are *not* formally named, so long will information about them be buried in the literature, where it is not available to botanists working in the field, who have to depend for detailed taxonomic information on Floras and monographs.

The point can be illustrated by reference to variation in *Viola riviniana* Rchb. (Valentine, 1941). In this paper a larger woodland and a smaller moorland ecotype were recognized, with ecoclinal connections between them, the smaller being named subsp. *minor* (Gregory) Valentine. The result was that authors of subsequent floras and papers were made aware of this variant (really first noted in Britain by Gregory, 1912), and have recorded it usefully in a number of localities.

More recently, in *Flora Europaea* (1968), subsp. *minor* has been brought down to the rank of variety (Valentine, 1968), as it is now held that the rank of subspecies is best reserved for geographical rather than ecological races. In addition, Valentine (1975) proposed to recognize and name a polymorphic character, at the rank of *forma*, based on the presence or absence of adventitious shoots on the roots. This is mentioned here, partly because the characters on which the varieties and the forms are based are both of ecological significance, and partly because it demonstrates that the taxonomic system can bear a reasonable load of information below the subspecific level. There is little doubt that the treatment proposed for *V. riviniana* is workable in western Europe; whether it will apply to the eastern part of the range of the species it is difficult, at present, to say.

It should be added that this system does not deal, by any means, with all the variation recorded in *V. riviniana*, or even with all the genotypic variation. There is almost certainly some change in leaf shape associated with geographical distribution, and, in addition, if local populations of *V. riviniana* are compared, they frequently differ, in an unsystematic way, in minor characters of indumentum, seed shape and weight, habit, colour of spur etc. To deal with all this information in formal terms would be very difficult; but to deal with the ecologically significant variation, in the way suggested, can be useful.

That the solution proposed above for *V. riviniana* is not exclusive can best be shown by quoting two other pieces of work which have been published in recent years. The first is that of Kay (1972) on *Tripleurospermum maritimum* (L.) Koch. Kay, in the full knowledge that the classification he is about to propose "to some extent . . . superimposes a discontinuous classification on a pattern of variation which is essentially continuous" realizes the potential value of a series of reference points, and after separating *T. maritimum* as a species from *T. inodorum* (L.) Schultz Bip. recognizes two subspecies, subsp. *maritimum* and subsp. *phaeocephalum* (Rupr.) Hämet-Ahti, with significantly different geographical distribution, the former in north and north-west Europe, the latter in subarctic and arctic Europe, extending to Greenland and North America.

He then proposes two varieties of subsp. *maritimum*, var. *maritimum* and var. *salinum* (Wallr.) Kay, distinguishable morphologically and with different though overlapping distributions, but not, apparently, ecologically different. This treatment thus differs from our proposal, in that ecological factors play little or no part at the infraspecific level.

The second example is that of Smith (1963) on *Melampyrum pratense* L. Here again is a very variable species, and Smith summarizes the pattern as follows:

There is geographical variation brought about by climatic selection from a basic pattern. Superimposed on this geographical variation is ecological variation caused by selection in different habitats. The third level of variation is brought about by the random isolation of particular genotypes in individual populations, leading to minor but detectable differences between populations in similar ecological habitats.

Smith discusses his extensive data in these terms, describing among other features a north-south cline in height and leaf-shape. He realizes the value of recognizing the significant variation, as far as possible, in taxonomic terms, and he comes to the conclusion that the most useful practical treatment is to recognize two subspecies based on ecological characters. These are subsp. *pratense*, found on acid soils, and subsp. *commutatum* (Tausch.) C. E. Britton, restricted to calcareous or base-rich habitats. The two subspecies differ in a series of morphological characters; they are both widely distributed in the British Isles and can also be recognized on the Continent. Smith retains var. *hians* Druce as a variety of subsp. *pratense*, characterized by flower colour and distinct geographical distribution, and he recognizes sporadic plants with crimson-tipped corollas as forma *purpureum* ined.

Again, Smith's treatment, like that of Kay, differs from that proposed for *Viola riviniana*; and this would point to the conclusion that, provided the treatment of the infraspecific variation is well

B

thought out and workable, it is best not to lay down rules about the use of subspecies and variety which are too rigid. The pragmatic approach is best; the main thing is to attempt to produce at least a taxonomic skeleton, which will guide others, such as students of ecology and evolution, who need to know what is going on. To what extent this system can be applied to widespread species, on a continental scale, it is difficult to say. Indeed, Böcher (1963), on the basis of an extensive knowledge of variation, expresses the view very strongly that in wide-ranging and variable species such as *Campanula rotundifolia* L., the pattern of variation is so intricate that it will never be possible to express it in taxonomic terms. Nevertheless, he admits that a limited number of variants can often be so treated; and it is my thesis that it is advantageous to the users of taxonomy to do this whenever it is possible.

As regards phenotypic variation which does not have a genotypic component, the taxonomic treatment is debatable. Often the assignment of formal names leads only to confusion. A case in point is *Plantago major* L. Turrill (1948) was able to show that this species was both tolerant and plastic. It could survive, in an experimental garden, on soils of a wide variety of textures, from sand to clay, and its size and form varied as it passed from one soil to another, being very small on sand and very large on clay. Such phenotypic plasticity and tolerance are not always recorded by taxonomists, or if so, under special terms such as ecads, outside the taxonomic hierarchy, and this is doubtless the best way. To burden the literature with a host of subsidiary taxa, without experimental investigation, can serve no useful purpose.

The habitat

The habitat can be characterized by many physical features, some climatic, some edaphic. In the former, details are rarely available for individual taxa, apart from a few which have been studied under experimental conditions, although it is sometimes possible to draw inferences from the coincidence of limits of distribution and climatic boundaries, such as lines of mean winter or summer temperature. More commonly, altitude limits are included in habitat descriptions, together with notes on such features as aspect and slope. Edaphic data are commonly available in some detail, and the chemical and physical characters of the soil (acidity, base status, texture, soil profile) can sometimes be used as indicators of the habitat preferences of the taxa concerned. Often in Floras, all these features, climatic and edaphic, are summed up in a brief phrase such as "maritime dunes" or "montane grassland", and these may be sufficient to correlate with taxonomic differences, especially where species are concerned.

It is possible to approach the problem from a different angle, and to characterize the habitat in terms of the kind of plant community which it supports. In Britain, though plant communities are well studied and often well described, they have not, until recently, been classified in a formal way; but on the continent of Europe, elaborate phytosociological classifications have been worked out, which have been widely applied to problems of descriptive ecology and land use. They have also been used to a varying extent by taxonomists, and there are several Floras which present a phytosociological classification of the plant communities in the area concerned, and assign each species of the Flora, as it is described, to its appropriate place in the classification. Perhaps the best example of its kind is the Flora of S. Germany by Oberdorfer (1962), in which the species are keyed out in the traditional way, but the ecology of each is given in detail, together with the phytosociological data. Valuable as this is, it does not represent an explicit contribution of ecological characters to taxonomy, and a more direct approach has been made by Guinochet (1955) and his school in France. Guinochet's ideas may be seen as a development of Turesson's ecotype concept but in more precise terms. On the ecological side, the habitat is defined in terms of modern phytosociology; on the taxonomic side, cytology is taken into account, and the data are often analysed by numerical methods. Guinochet's thesis is that the classification of plant communities in phytosociological terms presents a summing up of all the factors of the habitat, especially those physiological factors (including climate and soil) which are often hard to measure with precision. Thus, by comparing the occurrence of a particular taxon which occurs in two or more phytosociological units (e.g. associations), there is a *prima facie* case for inferring that the populations in these units are ecotypically distinct (whether they are morphologically distinct or not). The very fact of the occurrences stimulates the taxonomist to look for differences, (e.g. chromosomal differences) and if possible to do cultivation or growth chamber tests.

Using this approach, particularly in the montane and alpine areas of southern France, Guinochet and his pupils have been able to demonstrate the usefulness of phytosociology in disentangling and analysing taxonomically the populations of variable and critical species. Thus, in his studies of *Festuca ovina* sensu lato in the Alpes Maritimes, Bidault (1968) based his approach primarily on the formal and detailed account of the group by Saint-Yves (1913) in which taxa at the level of variety and even subvariety were recognized. He assessed the taxonomic data against the background of his detailed phytosociological knowledge of the communities concerned, and he then developed his investigation, using biosystematic techniques such as chromosome analysis and cultivation experiments. In his final taxonomic synthesis, which described the taxa mainly in terms of species and subspecies, all the relevant data were taken into account, and the phytosociological information finds its place in the descriptions of the taxa and the keys. It is probable that this method works best in the relatively undisturbed upland areas in which most of this kind of work has been done. Nevertheless, it represents an important contribution to synthetic taxonomy.

It is noteworthy that in many cases, not only in the work of the Guinochet school, but also in that of other continental workers, e.g. Küpffer (1974), the coincidence of diploid–polyploid boundaries with edaphic boundaries is observed. This is interesting in itself, as presenting a problem in mechanisms of adaptation, but it also raises the question of how to describe these chromosome races, often very difficult to define morphologically. Guinochet would use the phytosociological data in the taxonomic classification, as diagnostic characters. A similar procedure, using edaphic rather than vegetational differences, is used fairly widely, as for example in the ferns. Thus, in *Asplenium trichomanes* L., diploid and tetraploid populations are difficult to distinguish morphologically, and the fact that the diploids occur on base-poor and the tetraploids on base-rich rocks is significant. Taxonomists tend to distinguish such populations at the subspecific rather than the varietal level, the difference in chromosome number, because of its effect on genetical isolation, being given considerable weight (Crabbe et al., 1964). This practice is widespread.

THE ROLE OF ECOLOGICALLY INDUCED VARIATION IN THE ORIGIN OF SPECIES

Species as normally recognized by the taxonomist are defined in terms of (1) the mutual resemblance between the members of the populations concerned and (2) the degree of distinctness of one set of populations from another.

It is usual to think of some kind of initial isolation which is involved in the primary differentiation of species, and geographical isolation is one important mode. If and when the species subsequently come together and become at least partly sympatric, then some form of isolation must persist if they are to maintain their identity. This isolation can be of many kinds. It may be connected with pollinators or the breeding system; it may be chromosomal, as by polyploidy; or it may be ecological, in the sense that the species occupy different habitats.

Ecological isolation of this kind is not uncommon and, where it exists, the species can be characterized to some extent by their habitat differences. Examples of such pairs or groups of species in the British Isles have been given by Valentine (1951) under the name of gradual-ecospecies (a modification of Turesson's terminology). This and other specialized terminologies have not found favour in the eyes of taxonomists, who prefer the rock-like simplicity of their classical system; but in neglecting these terms (admittedly sometimes clumsy) they often lose sight of their important geographical and evolutionary implications.

Among the pairs of gradual-ecospecies which were mentioned by Valentine are the following:

Geum urbanum and *G. rivale*	$2n = 42$
Primula vulgaris Huds. and *P. veris* L.	$2n = 22$
Quercus robur L. and *Q. petraea* (Matt.) Liebl.	$2n = 24$
Epilobium hirsutum L. and *E. parviflorum* Schreb.	$2n = 36$
Silene maritima With. and *S. vulgaris* (Moench.) Garcke	$2n = 24$
Silene alba (Mill.) Krause and *S. dioica* (L.) Clairv.	$2n = 24$
Viola odorata L. and *V. hirta* L.	$2n = 20$

It will be noted that each pair is homoploid; the name gradual-ecospecies is given to them on the hypothesis that they evolved gradually by gene mutation and recombination without the intervention of polyploidy. At present, each pair is partly sympatric.

Where they meet, they do so at an ecological boundary or ecotone, and, though hybrids may be formed, the number and extent of these hybrids is limited by the spatial separation caused by the difference in habitat. (There are, of course, in most cases, other isolating factors as well, such as a difference in flowering time). Thus *Silene maritima* is usually a seashore and *S. vulgaris* an inland plant, *Primula vulgaris* usually a shade and *P. veris* a sun plant, *Geum urbanum* a plant of drier and *G. rivale* a plant of wetter habitats, and *Quercus robur* a tree of heavier and *Q. petraea* a tree of lighter soils. In no case except that of *Quercus* have the habitat differences between the species been at all precisely defined.

This pattern of speciation and isolation is common in temperate floras and in many genera. It is widespread in both *Quercus* and *Epilobium* and occurs also in such important genera as *Pinus* and *Salix*. In the sense that the species concerned are often differentiated geographically and often form partially fertile hybrids when they meet, the members of the group are often called subspecies (as for *Silene maritima* and *S. vulgaris* in *Flora Europaea*); but this is a matter of opinion. What is perhaps more important is that the meeting and the hybridization may set the scene for further evolutionary developments, either with or without polyploidy, and hybridization of this kind has undoubtedly been important in the evolution of both seed plants and ferns.

One evolutionary consequence of hybridization is introgression (Anderson, 1949). This begins at an ecotone, where species with different ecological requirements meet. The ecotone itself is occupied by F_1 hybrids, which may cross *inter se* or backcross to the parents, according to conditions. If the ecotone is narrow, and conditions undisturbed, then the hybrids are limited to the ecotone and do not spread. A good example of this is provided by the genus *Eucalyptus* in Australia, where there are many gradual-ecospecies, all with $2n = 22$. Pryor (1976) cites the case of *E. pauciflora* Sieber, which is found on granite, and *E. dives* Schau., found on slate. The two species are often found side by side at a narrow ecotone, and a few interspecific hybrids may be found. But the hybrids, though vigorous, are restricted to the ecotone, and do not compete successfully with either parent. If, however, the ecotone is broad, then the hybrids may become numerous; and if the habitat is disturbed, producing artificial conditions on one side of the ecotone or the other, then the hybrids may often prove to be as well or better adapted to the artificial conditions as the parents themselves. Under such conditions, genes may move from one parent to another, and the place of one parent be taken by backcross plants. An early and interesting example of this was described by Marsden-Jones (1930) in a hybrid population involving *Geum urbanum* and *G. rivale*.

Under other circumstances, hybridization at an ecotone may be followed sooner or later by polyploidy and the production of a new population genetically isolated from its parents and ecologically more or less intermediate. Again the rate of migration of the newcomer from its area of formation will depend on the nature of the ecotone and the habitats concerned. A famous example is that of the meeting of the North American *Spartina alterniflora* Lois. with the native *S. maritima* (Curt.) Fernald in Southampton Water in 1870, and the establishment first of the undoubled and then of the doubled polyploid hybrid, long known as *S. townsendii* H. & J. Groves. This proved to be pre-adapted to an ecological niche, in deep tidal mud, which was not occupied by either parent, whether native or introduced. In general, in spite of the wide ecological tolerance of many of the species involved in polyploid formation, the allopolyploids produced, judged by their abundance in the north temperate flora, must often have found vacant niches at their disposal.

The extent to which species of large genera are ecologically diversified is variable; it is a subject which has apparently not been investigated systematically. Sometimes, it is clear, the ecological character tends to belong to the genus, and the species pattern is geographically developed. This is broadly true of the rostrate section of the genus *Viola*, described by Valentine (1976). Again, as Strid (1970) has shown, one section of the genus *Nigella* has a series of endemic species on a number of Aegean islands which seem to be almost identical as regards habitat preference.

It will perhaps be of interest to conclude the chapter by referring to two genera whose ecological diversity has been well studied. The first of these is *Argyranthemum*, a genus of 22 species in the Compositae, 18 of which are centred on the Canary Islands (Humphries, 1976). The genus is of probable Early Tertiary origin, and can be divided into five distinct sections on the basis of cypsela morphology. The distribution of the five sections is closely correlated with the main vegetational zones, and each species consists of a series of discrete populations. All the species are diploid, and evolution has apparently been by gradual speciation, adaptation having taken place in a series of steep ecological gradients. Effective isolation is eco-geographical and hybrids are formed only in areas of considerable disturbance. The habitats are very diverse, ranging from coastal cliffs, sand-dunes and laurel scrub, to pine forest and mountain habitats. *Argyranthemum* is an epibiotic, derived from a palaeoendemic stock. Clearly ecological factors have played an important part in its recent evolution, and the ecological characters have a certain taxonomic importance.

The second example is *Epilobium* section *Epilobium* in New Zealand, which has recently been monographed by Raven and Raven (1976). In a penetrating piece of work, the authors trace the origin of the stock which migrated into New Zealand from Eurasia, directly or indirectly, in the Tertiary period, and its subsequent history and diversification. They show that, like its European relatives, it is diploid and inbreeding, but unlike these relatives, it is ecologically very diverse. The 37 native species occupy a very wide range of habitats, classified by Raven and Raven into 12 groups, and carefully characterized in terms of type of plant community, soil and altitude. Raven and Raven explain this diversification in terms of the ability of the species to form fertile hybrids and the inbreeding habit, which favours the development and stabilization of the diverse hybrid products, some of which find new ecological niches. In isolated oceanic islands such as New Zealand, genera capable of evolving in this way are likely to have had advantages in surviving the climatic and geological hazards of the Pleistocene era.

REFERENCES

Anderson, E. (1949). "Introgressive Hybridization". Wiley, New York.
Baker, H. G. (1956). *Geranium purpureum* Vill. and *G. robertianum* L. in the British Flora. II. *G. robertianum*. *Watsonia* 3, 270–279.
Bidault, M. (1968). Essai de taxonomie expérimentale et numérique sur *Festuca ovina* L., s.l., dans le sud-est de la France. *Rev. Cytol. et Biol. veg.* 31, 217–356.
Böcher, T. W. (1963). The study of ecotypical variation in relation to experimental morphology. *Regn. Veget.* 27, 10–16.

Briggs, D. and Walters, S. M. (1969). "Plant Variation and Evolution". McGraw Hill, New York.

Clausen, J., Keck, D. D. and Hiesey, W. M. (1940). Experimental studies on the nature of species. 1. The effect of varied environments on western North American plants. *Publs Carnegie Instn* **520.**

Crabbe, J. A., Jermy, A. C. and Lovis, J. D. (1964). *Asplenium. In* "Flora Europaea" (T. G. Tutin *et al.*, eds), Vol 1, pp. 14–17. Cambridge University Press, Cambridge.

Gilmour, J. S. L. and Gregor, J. W. (1939). Demes. A suggested new terminology. *Nature, Lond.* **144**, 333–334.

Gregor, J. W. (1938). Experimental taxonomy. 2. Initial population differentiation in *Plantago maritima* in Britain. *New Phytol.* **37**, 15–49.

Gregor, J. W. (1946). Ecotypic differentiation. *New Phytol.* **45**, 254–270.

Gregory, E. S. (1912). "British Violets". Heffer and Sons, Cambridge.

Guinochet, M. (1955). "Logique et Dynamique du Peuplement Végétal". Masson, Paris.

Humphries, C. J. (1976). Evolution and endemism in *Argyranthemum. Botanica Macaronesica* **1**, 25–50.

Huxley, J. S. (1938). Clines: an auxiliary taxonomic principle. *Nature, Lond.* **142**, 219–220.

Kay, Q. O. N. (1972). Variation in Sea Mayweed *Tripleurospermum maritimum* (L.) Koch in the British Isles. *Watsonia* **9**, 81–107.

Kerner von Marilaun, A. (1895) "The Natural History of Plants", Vol. 2. Blackie and Son, London.

Kruckeberg, A. R. (1954). The ecology of serpentine soils. III. Plant species in relation to serpentine soils. *Ecology* **35**, 267–274.

Küpffer, P. (1974). Recherches sur les liens de parenté entre la flore orophile des Alpes et celle des Pyrénées. *Boissiera* **23**, 11–322.

Marsden-Jones, E. M. (1930). The genetics of *Geum intermedium* Willd. haud Ehrh., and its back-cross. *J. Genet.* **23**, 337–395.

McNeilly, T. (1968). Evolution in closely adjacent plant populations. III. *Agrostis tenuis* in a small copper mine. *Heredity* **23**, 99–108.

Oberdorfer, E. (1962). "Exkursionsflora für Süddeutschland". Eugen Ulmer, Stuttgart.

Proctor, J. (1971). The plant ecology of serpentine. II. Plant response to serpentine soils. *J. Ecol.* **59**, 397–410.

Pryor, L. D. (1976). "Biology of Eucalypts". Edward Arnold, London.

Raven, P. H. and Raven, T. E. (1976). The genus *Epilobium* in Australasia. *New Zealand DSIR Bulletin* **216.** Christchurch.

Saint-Yves, A. (1913). Les *Festuca* de la section *Eu-Festuca* et leurs variations dans les Alpes maritimes. *Ann. Conserv. et Jard. Bot. Genève.* **17**, 1–218.

Salisbury, E. J. (1940). Ecological aspects of plant taxonomy. *In* "The New Systematics" (J. Huxley, ed.), pp. 329–340. Clarendon, Oxford.

Smith, A. J. E. (1963). Variation in *Melampyrum pratense* L. *Watsonia* **5**, 336–367.

Strid, A. (1970). Studies in the Aegean flora. XVI. Biosystematics of the *Nigella arvensis* complex. *Op. bot. Soc. bot. Lund.* **28**, 1–169.

Turesson, G. (1922a). The species and the variety as ecological units. *Hereditas* **3**, 100–113.

Turesson, G. (1922b). The genotypical response of the plant species to habitat and climate. *Hereditas* **3**, 211–350.

Valentine, D. H. (1941). Variation in *Viola riviniana* Rchb. *New Phytol.* **40**, 189–209.

Valentine, D. H. (1951). Geographical distribution and isolation in some British ecospecies. *In* "The Study of the Distribution of British Plants"(J. E. Lousley, ed.), pp. 82–90. Report of the Botanical Society of the British Isles Conference.

Valentine, D. H. (1968). *Viola. In* "Flora Europaea", (T. G. Tutin *et al*, eds.), Vol. 2, pp. 270–282, Cambridge University Press, Cambridge.

Valentine, D. H. (1975). The taxonomic treatment of polymorphic variation. *Watsonia* **10**, 385–390.

Valentine, D. H. (1976). Patterns of variation in north temperate taxa with a wide distribution. *Taxon* **25**, 225–231.

Wilkins, D. A. (1960). The measurement and genetical analysis of lead tolerance in *Festuca ovina. Scottish Plant Breeding Station Report* 85–98.

Chapter 2

Chemical Evidence in Plant Taxonomy

P. M. SMITH

INTRODUCTION

Plant taxa exhibit chemical variation which, like any other facet of their diversity, is theoretically a source of characters useful in taxonomy. Chemical evidence has, in fact, been used in this way ever since men first began to name and classify plants—edible and inedible categories being founded on chemical differences.

As knowledge of plant morphology and anatomy was extended by botanists, an awareness of chemical complexity also grew from investigations of the medicinal properties of plants and their extracts. Early information about plant chemistry, summarized in herbals, was largely concerned with the localization and application of physiologically active secondary metabolites such as saponins and alkaloids. During the eighteenth and nineteenth centuries this knowledge was greatly increased and certain aspects were used, by Eykman and Greshoff and others, in attempts to classify plants and to demonstrate their evolutionary relationships (Smith, 1976).

More and more kinds of natural plant products were gradually recognized, including proteins, nucleic acids and the major categories of polysaccharide. Greater insight into plant metabolism revealed an impressive underlying uniformity in the chemical structure and functioning of plants, but again drew attention to biochemical peculiarities and their possible taxonomic or evolutionary significance. Biochemical variation was scanned for discontinuities correlated with classification. This proved to be a rewarding field of enquiry, many claims being made for the taxonomic merit of various chemical characters. Thus Abbott (1886) suggested saponins might show particularly high predictive value, and Kowarski (1901) demonstrated that plant proteins certainly did.

In the present century much more has been learned about plant chemistry, most notably about the structure and function of nucleic acids and proteins. A general outcome has been a more informed appraisal of the meaning of discontinuities in the distribution of molecular types. We expect protein or nucleic variations to be taxonomically useful because we know that they reflect differences in the genetic specifications of the plants being compared. A richness in small molecules, such as amino acids and flavonoids, has been shown to be characteristic of plants. We know that some of this can be accounted for as part of the inevitable biochemical "noise" accompanying development, or the operation of the major metabolic pathways, but also that a substantial, taxonomically variable fraction cannot be so explained (Fraenkel, 1959).

Technical developments, particularly over the last 30 years, have made it possible to assay plant extracts in large numbers and comparatively quickly and cheaply. Previously the crudity and slowness of chemical extractions and analyses meant that it was difficult to apply them adequately to the study of plant variation, i.e. to the scrutiny of many different individuals within a taxon. Chromatography and electrophoresis have proved to be very suitable techniques for taxonomic surveys. Together with immunological techniques, which are most often applied to proteins, and molecular hybridization comparisons of nucleic acids, they comprise a category of "short cut" methods. These "short cuts" reveal a useful fraction of the information in the molecules, without resort to lengthy, expensive full-scale analysis. Even the current technology for automatic amino acid sequencing of proteins, a "total" form of analysis which has generated interesting facts and ideas on the phylogeny of both higher plants (Boulter et al., 1972) and bacteria (Ambler, 1972), seems unlikely to be applied routinely to taxonomic problems below the order level, for reasons of cost, material availability and time. Sequence analysis is not reviewed here because its specifically taxonomic contribution is marginal. In general it is now no more costly, time-consuming or difficult to search for chemical characters than to investigate any of the sources which botanists conventionally quarry for taxonomic evidence. Often the search is easier, quicker and cheaper.

Chemotaxonomic investigations have been rewarding at all hierarchical levels, from subvariety to division. Distributions of molecules as different as terpenoids and proteins have produced evidence of differential selection within populations. At the other extreme, the existence of cellulose, perhaps originally a photosynthetic waste product, around the cells of certain primitive organisms and their descendants helps us to distinguish them from animals. Another attraction of chemical characters is their generally broad homology, enabling very wide surveys of variation to be carried out, even across family boundaries. The search for chemotaxonomic affinities is not limited, as the biosystematic approach necessarily is, by the often narrow boundaries of possible gene exchange.

In a brief account such as this, only token examples of the applications of chemical evidence can be cited. A recent text (Smith, 1976) reviews the field comprehensively and provides copious references to chemotaxonomic literature. Encyclopaedias of chemotaxonomic data have been compiled by Hegnauer (1962–1969) and by Gibbs (1974). A series of volumes on the chemotaxonomy of particular flowering plant families (e.g. Vaughan *et al.*, 1976) is currently appearing. Greater attention to the functions of chemical characters has led to the discipline of chemical ecology on which books (e.g. Sondheimer and Simeone, 1970; Harborne, 1977) and a journal are now published.

THE NATURE OF CHEMICAL EVIDENCE

Chemotaxonomic facts are statements of the distribution of particular compounds. The occurrence of many simpler molecules in taxa can be expressed on a presence/absence basis. Thus ellagic acid is absent from all members of subfamily Rosoideae of Rosaceae, except for the single tribe Kerrieae (Bate–Smith, 1961). More complex molecules, such as proteins and nucleic acids, must be distributed in basically the same manner, but here it is usually harder to locate the boundaries between homologous types. This is because the molecules are rarely characterized fully by analysis. It has not so far been practicable to define a particular DNA sequence chemically so that its distribution can be investigated. Perhaps the possibility of hybridizing the DNA which codes for ribosomal RNA (rDNA) with its own transcription product—ribosomal RNA (rRNA)—will offer a feasible approach to such DNA characterization.

The well characterized proteins, such as cytochrome c, are virtually universal in distribution and can be compared fully only by amino acid sequencing. A combination of short-cut analyses, including serology, electrophoresis and testing separated proteins for enzyme function, provides the best opportunity for extracting taxonomic information from homologous proteins.

Taxonomic data from macromolecules hence tend to come in the form of patterns, ratios and percentages. Attempts have sometimes been made to transform them into other terms, e.g. the conversion of serological data into "immunological distance" (Prager and Wilson, 1971). It is important to facilitate the release of taxonomic evidence from these macromolecules because they contain a great deal of it. Zuckerkandl and Pauling (1965) term DNA, RNA and proteins, respectively, the primary, secondary and tertiary semantides—information-carrying molecules—of living organisms. Taxonomists are anxious to exploit them.

Chemotaxonomic evidence, whether in the form of presence/ absence data, patterns or other transformations, has been incorporated successfully into taximetric comparisons.

Taxonomic findings based on presence or absence of compounds are subject to change after further research. Wider sampling or more refined analysis may reveal the compound in taxa where it was previously unrecorded. Restricted distributions of a compound, and the taxonomic circumscriptions which may be founded on them, may turn out to be more apparent than real. Many free amino acids earlier thought to have restricted distributions were detected by Fowden (1972) in a bulk extract of sugar beet material. Nevertheless *accumulations* of particular amino acids or other substances may still be taxonomically meaningful, even though the biosynthetic machinery for their production may be widespread. Several of the free amino acids of legume seeds have interesting distributions. An example is canavanine, which seems not to occur outside the Papilionaceae.

Most kinds of compounds have been found to have taxonomic

significance in some taxa if not in others. Monosaccharide sugars are mostly so widely distributed that they display no interesting taxonomic variation. But rare monosaccharides exist. Acofriose, for example, is known only from *Acokanthera friesiorum* Markgraf (Apocynaceae). Sugar moieties of glycosides are likely to prove important taxonomic characters. Often a compound which is uniformly distributed at one level of the hierarchy is interestingly discontinuous at another. Structural and storage polysaccharides vary at the division level but, except for fungi and bacteria, not much below it. Ellagitannins are widely distributed but can be used to illuminate evolutionary theories in dicotyledons (Sporne, 1975).

Chemotaxonomists have discovered merit in the time-honoured principle of considering as many characters as they can. Data from protein electrophoresis are often usefully supplemented by serological enquiries (Pickering and Fairbrothers, 1967). Multiplicities of chemical characters are studied in bacterial taxonomy—evidence from metabolic output and wall chemistry complementing information on DNA base ratios and hybridization. Distribution of several kinds of secondary substances are investigated increasingly in parallel. Information on amino acid sequences can be considered together with immunological comparisons of the same proteins (Wallace and Boulter, 1976).

The taxonomic value of particular kinds of chemical evidence varies according to the considerations which apply to all phenetic characters. Factors which determine the value of a character include its stability over a range of environmental conditions, its constant association with a number of other features, i.e. its predictive quality, its complexity and functional significance, and the ease with which it can be detected. Some of these factors are reviewed briefly below.

Environmentally induced variation

Micromolecules such as amino acids, phenolics and alkaloids have been thought to be subject to considerable environmentally induced variation, as well as to change during the course of plant development. Both quantitative and qualitative changes can indeed be shown to arise from both these sources, but so long as their existence and magnitude are recognized they need not detract from the taxonomic usefulness of a character. Protein characters have often been considered more stable, because they are effectively buffered from non-selective environmental influences. But is it now axiomatic that protein complements change considerably during development. Environmental changes can sometimes have direct effects on protein characters (Bradford *et al.*, 1975; Blagrove *et al.*, 1976). Proteins from homologous organs of the same physiological age offer a sound basis for taxonomic comparison.

Correlation with other features

Chemical characters of many types have often been found to be correlated with each other, and with other kinds of data as diverse as crossability (Gell *et al.*, 1960; Pandey, 1967; Smith, 1972), genome relationships (Johnson and Hall, 1965) and grafting compatibility (Rives, 1923; Kloz, 1971). Overall correlation with the commonplace yet cardinal data of morphology and anatomy is of a high order.

Complexity and taxonomic value

The complexity of a macromolecule character arises from the number of steps in its pathway of synthesis as well as from its eventual structure. This complexity reduces the possibility of the character being shared by chance between a number of taxa. Convergent evolution is unlikely to have occurred because the character depends on so many genes. Its taxonomic value thus rests partly on its complexity, and is high.

On the other hand, a simple molecule is *per se* a simple character. Its mere incidence may have no taxonomic implications. Only if the distribution of the molecule is likely to be the same as that of a pathway of synthesis, and the genes which control it, can the simple molecule be taxonomically predictive, i.e. of high value. Though the molecule itself is a simple character, constant association with a complex synthetic pathway can lend it complexity. Few simple molecules have been reliably associated with a constant biosynthetic pathway. Some simple molecules occur very sporadically in the plant kingdom. For example, the alkaloid nicotine occurs in several unrelated orders, in genera ranging from *Nicotiana* to *Equisetum*. It is highly likely that it is produced by several different pathways, and that its general occurrence is taxonomically meaningless. A way in which a simple molecule, of unknown biosynthetic origin, may gain taxonomic value is of course where it correlates with other chemical characters. Alkaloid characters often occur in associated groups— where one alkaloid is present others are normally found also.

Function and taxonomic value

Characters to which no demonstrable or speculative function can be assigned are regarded suspiciously by taxonomists. Possibly they have arisen by chance, in which case the plants which exhibit them need have no close relationship. Function implies selection. Selection underlies taxonomic divergence.

Macromolecular characters are associated with obvious, important functions in metabolism, storage, differentiation and growth. Secondary metabolite characters have been looked at somewhat more askance. This has been partly because of the suspicion that they are accidental or waste products without functional significance. In recent years there has been long overdue research into the possible functions of these substances and we now know quite a lot about them. It seems likely that most of them will prove to have considerable selective advantages in reducing biotic threats by acting, for example, as grazing deterrents, antibiotics or phytoalexins. Production of allelopathic compounds creates a useful biotic threat. Greenhalgh and Mitchell (1976) recently demonstrated that the flavour volatiles characteristic of *Brassica oleracea* L. confer resistance to powdery mildew. Such functions undoubtedly increase the value which taxonomists will in future place on discontinuities in the distribution of secondary substances.

To summarize, so long as chemical characters are considered as critically as other phenetic evidence, and the ground rules of taxonomic enquiries are obeyed, chemotaxonomic contributions will be real and valuable. The best way to introduce the scale of the contributions is to assess some different examples.

CHEMICAL EVIDENCE AT FAMILY AND ORDER LEVELS

Above the level of the family, certain chemical characters assist the fundamental classification of plants. Some of these have already been cited. Studies of serological relationship, based on protein comparisons, have frequently transgressed family boundaries, providing useful information about the size of interfamily discontinuities [e.g. Jensen (1968) on Ranunculaceae; Fairbrothers and Johnson (1964) on Cornaceae/Nyssaceae]. A few secondary metabolite distributions have proved helpful guides to the delimitation of families [e.g. ranunculin in Ranunculaceae (Ruijgrok, 1966)].

The value of betalain distribution in the definition of the order Centrospermae (Caryophyllales) is the best-known example of chemical evidence resolving taxonomic problems at order level. The betalain compounds (betacyanins and betaxanthins) replace the widespread anthocyanins and anthoxanthins as the red, blue and yellow pigments in most families belonging to the Centrospermae. Table I shows a comparison of the traditional circumscription of the Centrospermae, based chiefly on anatomy and morphology, with that of families which possess betalain pigments.

TABLE I. *Three classifications of the Centrospermae and their relatives.*

Orthodox	Chemical	Behnke and Turner (1971)
Centrospermae	**Centrospermae**	**Caryophyllidae**
Molluginaceae	Aizoaceae[a]	*Caryophyllales*
Caryophyllaceae	Portulacaceae[a]	Caryophyllaceae
Aizoaceae[a]	Phytolaccaceae[a]	Molluginaceae
Portulacaceae[a]	Chenopodiaceae[a]	
Phytolaccaceae[a]	Amaranthaceae[a]	*Chenopodiales*
Chenopodiaceae[a]	Didiereaceae[a]	Phytolaccaceae[a]
Amaranthaceae[a]	Nyctaginaceae[a]	Nyctaginaceae[a]
Didiereaceae[a]	Basellaceae[a]	Didiereaceae[a]
Nyctaginaceae[a]	Cactaceae[a]	Cactaceae[a]
Basellaceae[a]		Aizoaceae[a]
	Excluded	Portulacaceae[a]
Cactales	Molluginaceae	Basellaceae[a]
Cactaceae[a]	Caryophyllaceae	Chenopodiaceae[a]
		Amaranthaceae[a]

[a]Families in which betalains have been recorded.

It will be seen that, in the "chemical" Centrospermae, the Cactaceae are included. Cactaceae have sometimes been suspected of centrospermous affinities partly because of their vegetative and floral similarities with Aizoaceae. Most taxonomists have nevertheless considered these similarities insufficient to warrant the inclusion of the Cactaceae in the Centrospermae, perhaps believing they were due to convergent evolution in arid environments. So the Cactaceae have usually been placed, rather unsatisfactorily, in a monotypic order of their own (Cactales or Opuntiales). The Caryophyllaceae, which on orthodox criteria are clearly of centrospermous affinity, are cast out of the order on the chemical evidence, since they do not contain betalains. Molluginaceae also are excluded. This clash of orthodox and chemical evidence is augmented by Behnke's discovery of characteristic protein inclusions in the sieve-tube plastids of the plants in the Centrospermae more or less as traditionally defined. This ultrastructural evidence suggests that Cactaceae, Caryophyllaceae and Molluginaceae are correctly assigned to the Centrospermae. A compromise classification (Table I) has been proposed which in essence replaces the Centrospermae by two very closely related orders—the Caryophyllales *sensu stricto* (without betalains) and the Chenopodiales (with betalains), but includes them both in the subclass Caryophyllidae (Behnke and Turner, 1971).

CHEMICAL EVIDENCE AT GENERIC AND SPECIFIC LEVELS

Many chemotaxonomic contributions have concerned generic relationships. Several are reviewed by Smith (1976). Among the most recent is a flavonoid survey of Ericaceous genera in which strong support is shown for a new anatomically based classification (Harborne and Williams, 1973).

Taxonomists of bacteria have for many years been compelled to rely on chemical characters for basic classification and identification. They have had few other sources of evidence. Newer chemotaxonomic methods have been enthusiastically applied to problems of microbial systematics and they have frequently offered new insights (Smith, 1976). Recent work on the actinomycete genus *Bacterionema* (Alshamaony *et al.*, 1977) compared whole-organism methanolysates. These were found to contain mycolic acids— substances never before isolated from actinomycetes. In structure and size, the molecules resemble those typically found in *Corynebacterium*. Hence the workers examined the merit of transferring *Bacterionema* organisms to the corynebacteria. Other evidence also supported this transference: the long chain fatty acids of *Bacterionema* were similar to those of *C. diphtheriae*. Earlier evidence from wall compounds and pathways of propionic acid fermentation also indicates a similarity between coryneform bacteria and *Bacterionema*. DNA base composition (the guanine-cytosine percentage) further supports the idea of this relationship. Alshamaony *et al.* (1977) regard it as demonstrated. It is interesting that *Bacterionema* began its nomenclatural life as a monotypic repository for three species of oral micro-organisms, which were formerly referred to two other genera. The transfer now proposed to the corynebacteria is a simplification which opens the way for a further study of the precise affinities of the taxon to other species of *Corynebacterium*. Chemotaxonomic evidence is encouraging many rationalizations of this kind in a group of organisms where taxonomic clarification is keenly welcomed.

Flavonoid variation is probably the facet of chemical diversity easiest to expose, and for this reason it has been most widely applied in taxonomic studies. Recent work on liverwort taxonomy has shown at once how broadly applicable, how simple and how productive such studies can be. One example concerns a 50-year-old controversy over how many species can be recognized in the genus *Hymenophyton* (Markham *et al.*, 1976). Rapid flavonoid extraction, chromatographic analysis and identification have showed consistently different patterns for two entities which some taxonomists have accepted as species, but which others have lumped together. Figure 1 shows the patterns which were obtained. The chemotaxonomic data support the argument that *Hymenophyton* is not monotypic, but that *H. leptopodum* (Labill.) Dumort. can be recognized as a species distinct from *H. flabellatum* Steph. Here chemical evidence is helpful in supporting one taxonomic interpretation of morphological data, but not the other. As often happens in such situations, the chemical findings stimulated a careful re-examination of the morphological evidence, which has revealed inadequacies in the original observations. A further valuable feature of this very simple investigation is that it marks the first demonstration of flavonoids in the Metzgeriales, and may indicate that, as some bryologists have

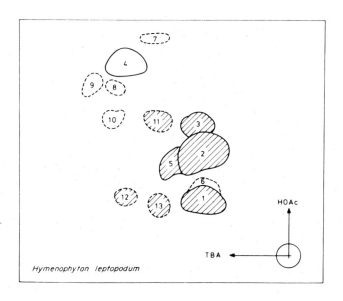

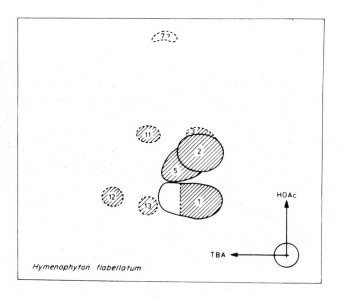

Fig. 1. Two-dimensional paper chromatograms of the flavonoids in *Hymenophyton leptopodum* and *H. flabellatum*. Shaded spots are common to both species. Faint spots are represented by dotted outlines. From Markham *et al.* (1976).

supposed, *Hymenophyton* is an advanced group, having close phylogenetic relationships with the Marchantiales.

Serological techniques are being used to provide one of several lines of new taxonomic evidence in a revision of the large grass genus *Bromus* (Smith, 1971, 1972). Figure 2 shows some immuno-electrophoretic separations of seed globulins of perennial species of this genus. They form part of the data being accumulated in a comparison of Old World and New World members of the section Pnigma. The methods used to produce the evidence have already been described (Smith, 1972). Only the major proteins are compared by this technique. Figure 2(a) shows some reactions of an antiserum raised to *B. ciliatus* L., a widespread North American diploid species, to a range of seed extracts (antigens) from other taxa. The antiserum is able to recognize, and precipitate efficiently, more protein antigens in a *B. ciliatus* extract than any other, as might be expected. *B. ciliatus* antigens produce the "homologous" reaction to that antiserum, the others giving "heterologous" reactions. The heterologous reactions show a fair degree of variation among the other species tested—*B. anomalus* Rupr. ex Fourn. being most similar to *B. ciliatus*. *B. orcuttianus* Vasey seems to share many proteins in common with *B. ciliatus*, but the electrophoretic mobility of most of them is different. Of the six spectra of diploids illustrated, *B. vulgaris* (Hook.) Shear is most different from *B. ciliatus* in numbers of reacting systems and in the electrical mobility of the antigenic proteins. Extracts of *B. pumpellianus* Scribn., a northern octoploid species with supposed Eurasian-North American evolutionary links, contain nine protein antigens which can regularly be precipitated by the *B. ciliatus* antiserum. On the face of it, it does not look *very* different in its immunoelectrophoretic pattern from the six diploid taxa.

Figure 2(b) shows reactions of some of the same extracts to antiserum raised to *B. pumpellianus*. This test reveals that *B. pumpellianus* antiserum is unable to resolve many differences between the North American diploid taxa—which produce very similar spectra—with the exception of *B. vulgaris*. A fair difference exists between the homologous reaction (*pumpellianus* antiserum/ *pumpellianus* antigens) and the heterologous reaction to proteins of *B. inermis* Leyss., a Eurasian octoploid, which of those reactions here illustrated is most similar to the homologous reaction.

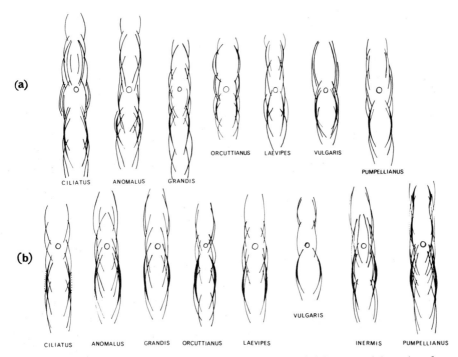

FIG. 2. Immunoelectrophoretic patterns of seed proteins of eight perennial species of *Bromus*. In each case the antigen origin is represented by the central circle. Antisera were applied in troughs parallel to the axis of electrophoretic separation. (a) Patterns produced by antiserum to *B. ciliatus*. (b) Patterns produced by antiserum to *B. pumpellianus*. For technique details, see Smith (1972).

The results of using the *B. ciliatus* antiserum, which are confirmed by using antisera to *B. laevipes* Shear and *B. orcuttianus*, show that the North American diploids are reasonably diverse. This could not have been predicted confidently from their morphology, which is reasonably uniform. The subtle morphological differences which exist would seem to be meaningful. Earlier comparisons with *B. pumpellianus* antiserum had suggested that the North American diploids, with the exception of *B. vulgaris*, were serologically very uniform (Fig. 2b), despite some interesting differences of geographical and ecological distribution. Using antisera to two very different species it is possible to take bearings, as it were, on the relative serological relationships of species from two points.

B. ciliatus $(2n = 14)$ probably shares a genome with *B.*

pumpellianus ($2n = 56$), so it is likely that *B. ciliatus* antiserum would detect a number of *B. pumpellianus* proteins. The reverse is also likely, and demonstrated by these data. Eight or nine of their seed proteins are very similar serologically and electrophoretically. But *B. pumpellianus* antiserum is unable to resolve the finer differences between the diploids—which presumably arose during their ancient adaptive radiation in North America. It reveals a fair number of differences between the homologous reaction and that of the nearly related species *B. inermis* ($2n = 56$). Antisera to *B. laevipes, B. ciliatus* and *B. orcuttianus* do not distinguish very clearly between *B. inermis* and *B. pumpellianus* (spectra not shown).

Antisera therefore give a clear indication of differences between species related to the homologous species, but a coarser indication of differences between less related taxa. The coarser view enables the major groups of taxa to be recognized— in this restricted example some diploid species of western and central North America, and the boreal octoploids. The fine resolution exposes interspecific differences between closer taxa, which can be evaluated in concert with data from other sources. It is interesting that both antisera here mentioned have indicated an unexpectedly anomalous position for *B. vulgaris*.

Taking further antisera as guides it should be a simple matter to examine the interrelationships of other North American diploids, the numerous perennial taxa of Eurasia, and the curious taxa of Africa and South America. This task, covering 70 or 80 species, is proving to be interesting and rewarding in terms of varying degrees of correlation with other evidence—both chemical and non-chemical— and the occasional exciting discrepancy.

Some work on species relationships in *Chenopodium* (Crawford and Julian, 1976) illustrates the increasing trend to investigate more than one kind of chemical evidence at a time. In 69 populations of seven species of *Chenopodium*, an electrophoretic study of water-soluble seed proteins showed little or no intraspecific variation. A flavonoid survey of some taxa produced data which were sometimes fully compatible with interspecific protein differences, but which sometimes did not agree. *Chenopodium atrovirens* Rydberg and *C. leptophyllum* Nutt. ex Moq. had identical flavonoid patterns but could be distinguished by their different seed protein spectra. *C. desiccatum* A. Nelson and *C. atrovirens* were closely similar in seed proteins, but differed from each other in flavonoids, and in other characters. Both flavonoid and protein evidence enabled *C. hians* Standley to be distinguished from *C. leptophyllum*, and showed *C. atrovirens* to be very similar to *C. pratericola* Rydberg. It is hard to generalize about which of these two kinds of evidence should be weighted more heavily. Much depends on the strength of other evidence, and the degree of uncertainty felt by taxonomists in particular delimitations. *C. hians* and *C. leptophyllum* are controversial taxa—sometimes lumped together, sometimes kept as separate species. Here both flavonoid and protein data support the latter treatment. Where chemical findings clash, the wise course is to look for more evidence. There is no philosophical difficulty in explaining the meaning of clashes of different sorts of evidence. Where characters have vastly different functions, such as flavonoids and proteins, where they are likely to have evolved at different rates and responded differentially in phases of introgression, and where they react to different environmental pressures, it is unlikely that the discontinuities they show will coincide fully. This applies to all kinds of taxonomic evidence. It would be foolish to think that in chemical evidence we have the ultimate taxonomic guide. Some workers have in the past supposed that chemical variations are somewhat more fundamental than others and hence should *a priori* be heavily weighted by taxonomists. At other times taxonomists have been encouraged to genuflect with the same promptness before findings of anatomy, cytology and genetics.

CHEMICAL EVIDENCE AT THE POPULATION LEVEL

Chemical variation within populations is well known: it would be possible to cite examples of population variation in every kind of chemical character. The scale and taxonomic usefulness of this variation depends on the function of the character, the breeding system of the plant and any taxonomic problem which that creates. Smith (1976) discusses a number of cases where chemotaxonomic evidence is helpful in recognizing minor variants and in understanding the significance of intra- and interpopulation variability.

Sesquiterpene lactone characters have proved helpful in both population and hybridization studies in the Compositae. Work on *Xanthium strumarium* L. (McMillan *et al.*, 1976) exemplifies both. This species has populations in both the Old and New Worlds. Eight major sesquiterpene lactones were identified in leaf extracts by thin-layer chromatography or nuclear magnetic resonance spectroscopy. The inheritance of these was followed in F_1 hybrids produced between plants belonging to various New World morphological complexes, and *X. strumarium* populations indigenous to the Old World.

Old World populations produce xanthinin and/or xanthinosin with some related compounds. New World populations contain xanthinin or its stereoisomer xanthumin. Plants of the *chinense* complex (from S. Louisiana) contain xanthumin and are believed to be the probable source of introduced *chinense* populations in India and Australia. Some of the Indian plants were shown to produce a xanthumin derivative (diacetyloxyxanthumin) additional to xanthumin itself. Artificial F_1 hybrids of Old World *X. strumarium* plants with pure *chinense* plants reveal a similarly mixed sesquiterpenoid pattern. This suggests that natural introgression of indigenous Old World *X. strumarium* and introduced *chinense* has occurred. Some Old World *chinense* populations produce xanthinosin, which is absent from the Louisiana *chinense* plants but is common in Old World populations of *X. strumarium*. Again this points to a flow of genes between the two taxa in India.

It is not yet clear from these interesting sesquiterpenoid data whether the Old World populations gave rise to the American ones, or vice versa. The New World populations as a whole show greater chemical diversity, which might suggest that the native Old World *X. strumarium* plants represent a derived type. The basic stock, probably American on this theory, would have produced xanthinin as the major sesquiterpenoid, and it would have evolved into the xanthinosin-producing plants of Asia and into xanthumin-rich populations in other parts of America. On the other hand, it is possible that the original *X. strumarium* plants, probably Asian, produced only xanthinosin, like many Old World plants today. In this case the chemical diversification of New World plants must have accompanied their adaptive radiation into new habitats, after the original colonization. Much later they have exported some of the diversity to the Old World via introductions of the *chinense* type from the south-eastern United States. It is noteworthy that xanthinosin production is almost or quite suppressed in hybrids between Asiatic xanthinosin producers, and American xanthinin or xanthumin producers. Hence, although these chemical analyses allow hypotheses to be devised, very different sorts of evidence will be needed to decide between them. In population studies founded on the sesquiterpenoids of *Ambrosia* (Potter and Mabry, 1972) an evaluation of morphological, cytological and ecological considerations enabled plausible conclusions to be drawn from similar chemical results.

Chemical evidence, therefore, offers useful new facts on population relationships, just as previously seen at other hierarchical levels. In each case it is clear that chemical characters can be profitably employed only when intelligently correlated with all other evidence available. We judge them to be valuable, often critically important, but not *inherently* definitive. In taxonomy, all evidence is circumstantial.

FUTURE CHEMOTAXONOMIC DEVELOPMENTS

The application of chemical evidence to taxonomy is still in a phase of active expansion in which the scale of its contribution is not yet fully calculable. In order to realize its full potential contribution, a number of developments are necessary, some of which are already taking place. There must be greater chemical awareness on the part of taxonomists, and greater understanding of taxonomic and evolutionary problems by those who produce the raw chemical data. Biochemical short cuts must be improved and extended. Technological advance must be considerable before the genetic information of plants, encoded in nucleic acids, can be compared with classical taxonomic information. Chemical data must be presented in ways which will facilitate their incorporation into taxonomic monographs.

There must be more research into the functional significance of chemical characters, and the selection pressures which act on them. More work on the biosynthesis of secondary metabolites must be done to illuminate their taxonomic value. Taxonomic surveys must be widened to expose spuriously "restricted" distributions of compounds. The broader correlations of chemical characters must be examined; for example, the connections of protein variation with brccding barricrs and with grafting compatibility might reward investigation. Application of chemotaxonomic methods to more critical taxonomic problems would increase their productivity. Surveys of critical genera and of groups of heterogeneous, primitive and isolated taxa would be especially interesting. It is most important that future gatherers of chemotaxonomic evidence should not be just chemists or taxonomists *manqués*. Chemotaxonomic methods and facts must be properly assimilable into the general discipline before systematists can fully exploit the treasure house of chemical characters which is now open.

REFERENCES

Abbott, H. C. de S. (1886). Certain chemical constituents of plants considered in relation to their morphology and evolution. *Bot. Gaz.* **11**, 270–272.

Alshamaony, L., Goodfellow, M., Minnikin, D. E., Bowden, G. H. and Hardie, J. M. (1977). Fatty and mycolic acid composition of *Bacterionema matruchotii* and related organisms. *J. Gen. Microbiol.* **98**, 205–213.

Ambler, R. P. (1972). Sequence data acquisition for the study of phylogeny. *In* "Recent Developments in the Chemical Study of Protein Structure" (A. Previero, J.-F. Pechere and M. A. Colleti-Previero, eds), pp. 289–305. University of Montpellier, Montpellier.

Bate-Smith, E. C. (1961). Chromatography and taxonomy in the Rosaceae, with special reference to *Potentilla* and *Prunus*. *J. Linn. Soc. (Bot.)* **58**, 39–54.

Behnke, H. –D., and Turner, B. L. (1971). On specific sieve-tube plastids in Caryophyllaceae. *Taxon* **20**, 731–737.

Blagrove, R. J., Gillespie, J. M. and Randall, P. J. (1976). Effect of sulphur supply on the seed globulin composition of *Lupinus angustifolius Aust. J. Plant Physiol.* **3**, 173–184.

Boulter, D., Ramshaw, J. A. M., Thompson, E. W., Richardson, M. and Brown, R. H., (1972). A phylogeny of higher plants based on the amino acid sequence of cytochrome *c* and its biological implications. *Proc. R. Soc., B,* **181**, 441–455.

Bradford, L. S., Jones, R. J. and Garber, E. D. (1975). An electrophoretic survey of fourteen species of the fungal genus *Ustilago*. *Bot. Gaz.* **136**, 109–115.

Crawford, D. J. and Julian, E. A. (1976). Seed protein profiles in the narrow-leaved species of *Chenopodium* of the Western United States: Taxonomic value and comparison with distribution of flavonoid compounds. *Am. J. Bot.* **63**, 302–308.

Fairbrothers, D. E. and Johnson, M. A. (1964). Comparative serological studies within the families Cornaceae (Dogwood) and Nyssaceae (Sour Gum). *In* "Taxonomic Biochemistry and Serology" (C. A. Leone, ed.), pp. 305–318. Ronald Press, New York.

Fowden, L. (1972). Amino acid complement of plants. *Phytochemistry* **11**, 2271–2276.

Fraenkel, G. S. (1959). The raison d'etre of secondary plant substances. *Science, N.Y.* **129**, 1466–1470.

Gell, P. G. H., Hawkes, J. G. and Wright, S. T. C. (1960). The application of immunological methods to the taxonomy of species within the genus *Solanum. Proc. R. Soc., B.* **151**, 364–383.

Gibbs, R. D. (1974). "Chemotaxonomy of Flowering Plants," Vols 1–4. McGill—Queen's University Press, Montreal.

Harborne, J. B. (ed.) (1977). "Introduction to Ecological Phylogeny". Academic Press, London and New York.

Harborne, J. B. and Williams, C. A. (1973). A chemotaxonomic survey of flavonoids and simple phenols in leaves of the Ericaceae. *Bot. J. Linn. Soc.* **66**, 37–54.

Hegnauer, R. (1962–1969). "Chemotaxonomie der Pflanzen," Vols 1–5. Birkhauser, Basel.

Jensen, U. (1968). Serologische beitrage zur Systematik der Ranunculaceae. *Bot. Jb.* **88**, 204–268.

Johnson, B. L. and Hall, O. (1965). Analysis of phylogenetic affinities in the Triticinae by protein electrophoresis. *Am. J. Bot.* **52**, 506–513.

Kloz, J. (1971). Serology of the Leguminosae, *in* "Chemotaxonomy of the Leguminosae" (J. B. Harborne, D. Boulter and B. L. Turner, eds), pp. 309–365. Academic Press, London and New York.

Kowarski, A. (1901). Über den Nachweis von pflanzlichem Eiweiss auf biologischem Wege. *Dt. med. Wschr.* **27**, 442.

Markham, K. R., Porter, L. J., Campbell, E. O., Chopin, J. and Bouillant, M.-L. (1976). Phytochemical support for the existence of two species in the genus *Hymenophyton. Phytochemistry* **15**, 1517–1521.

McMillan, C., Mabry, T. J. and Chavez, P. I. (1976). Experimental hybridization of *Xanthium strumarium* (Compositae) from Asia and America. II. Sesquiterpene lactones of the F_1 hybrids. *Am. J. Bot.* **63**, 317–323.

Pandey, K. K. (1967). Origin of genetic variability: combination of peroxidase isozymes determine multiple allelism of the S gene. *Nature, Lond.* **213**, 669–672.

Pickering, J. L. and Fairbrothers, D. E. (1967). A serological and disc electrophoretic investigation of *Magnolia* taxa. *Bull. Torrey bot. Club* **94**, 468–479.

Potter, J. L. and Mabry, T. J. (1972). Origins of the Texas Gulf coast island populations of *Ambrosia psilostachya*: a numerical study using terpenoid data. *Phytochemistry* **11**, 715–723.

Prager, E. M. and Wilson, A. C. (1971). The dependence of immunological cross-reactivity upon sequence resemblence among lysozymes. I. Microcomplement fixation. *J. biol. Chem.* **246**, 5979–5989.

Rives, L. (1923). Sur l'emploi du serodiagnostic pour la determination de l'affinité au gréffage des hybrides de vigne. *C. r. hebd. Séanc. Acad. Agric. Fr.* **9**, 43–47.

Ruijgrok, H. W. L. (1966). The distribution of ranunculin and cyanogentic compounds in the Ranunculaceae In "Comparative Phytochemistry" (T. Swain, ed), pp. 175–186. Academic Press, London and New York.

Smith, P. M. (1971). The taxonomy and nomenclature of the brome-grasses. *Notes R. bot. Gdn. Edinb.* **30**, 361–375.

Smith, P. M. (1972). Serology and species relationships in annual bromes (*Bromus* L. sect. Bromus). *Ann. Bot.* **36**, 1–30.

Smith, P. M. (1976). "The Chemotaxonomy of Plants" Edward Arnold, London.

Sondheimer, E. and Simeone, J. B. (1970). "Chemical Ecology." Academic Press, London and New York.

Sporne, K. R. (1975). A note on ellagitannins as indicators of evolutionary status in dicotyledons. *New Phytol.* **75**, 613–618.

Vaughan, J. G., Macleod, A. J. and Jones, B. M. G. (eds). (1976). "The Biology and Chemistry of the Cruciferae." Academic Press, London and New York.

Wallace, D. G. and Boulter, D. (1976). Immunological comparisons of higher plant plastocyanins. *Phytochemistry* **15**, 137–141.

Zuckerkandl, E. and Pauling, L. (1965). Molecules as documents of evolutionary history. *J. theor. Biol.* **8**, 357–366.

Chapter 3

The Chromosomes and Plant Taxonomy

D. M. MOORE

INTRODUCTION

The discovery of chromosomes in the last quarter of the nineteenth century and the subsequent demonstration that they are the carriers of genetic information provided an impetus to, or at least a physical foundation for, the experimental evolutionary studies on which are based our knowledge of such taxonomically important matters as variation within plant species and the significance of the breeding structure of populations and of the barriers to gene exchange between them. These studies, with their emphasis on the importance of "gene-flow" and "genetic discontinuities" in delimiting the basic units of evolution, led to the view, propounded by many cyto-geneticists in the early decades of this century, that genetically circumscribed "biological species" were somehow more "real" than the morphologically delimited "classical" species defined more or less at the whim of the herbarium taxonomist.

Because of its long history, it is inevitable that the bases, categories and, in part, the methodology of taxonomy were laid down in the days prior to the publication of *The Origin of Species*. Furthermore, during the late nineteenth and into the twentieth centuries, the theory of evolution had little effect on taxonomic procedures or on the systems resulting from them, even though the taxonomist, who has traditionally described and ordered into a hierarchy the products of evolution on the basis of apparent similarity and dissimilarity, was not slow to appreciate its significance. However, the first 30 years of the present century saw an increasing impact of cytogenetical concepts and procedures on at least some taxonomists. The increased output of chromosome data from a wide variety of plants, facilitated by the development of the squash technique, and the experimental demonstration, in genera such as *Nicotiana*, of hybridization followed by polyploidy as a method of speciation, provided some of the fuel for the New Systematics (Huxley, 1940), by which the arbitrary and often imperfectly documented procedures and decisions of the "classical" or "herbarium" taxonomists were to be replaced by a certainty based on sound cytogenetical principles: "the chemists had the periodical table of the elements, why could taxonomists not have repeatable units also?" (Raven, 1974).

During these years, the important role of the chromosomes in aiding taxonomic understanding became evident through classical and much quoted studies on Cruciferae, Gramineae, Agavaceae, and later, incisive monographs of, for example, *Crepis*, *Datura*, *Nicotiana*, *Clarkia*, *Tragopogon* and pteridophytes. Biosystematics (Camp, 1951), or experimental taxonomy, which then as now was concerned largely with data derived from studies of the chromosomes, became the growing point of taxonomy, with Stebbins's (1950) masterly *Variation and Evolution in Plants* a powerful stimulus. The number, size and shape of chromosomes were used to characterize the karyotypes of plants and define the taxonomic differences between them; the pairing behaviour of the meiotic chromosomes in natural or, more usually, artificial hybrids between plants of different populations indicated the extent to which they had differentiated and, in some cases, showed how genomes had been combined like building blocks by polyploidy to provide new species. Support for ecospecies, coenospecies and comparia, delimited cytogenetically, reflected the search for a hierarchy of precisely defined taxonomic units.

During the 1960s, however, a mood of some disillusion began to creep into biosystematic discussions and publications. Cytogenetic procedures and concepts did not provide the universal panacea for taxonomic problems that some had forecast and the reasons were increasingly clear (e.g. Jones, 1970). Meiotic pairing sometimes reflected chromosomal homology, and sometimes resulted from the specific action of genes controlling synapsis (Riley, 1966; de Wet and Harlan, 1972). Surveys of haploidization (Raven and Thompson, 1964; de Wet, 1968; Khasha, 1974) cast doubt on the traditionally accepted relationship between diploids and derived polyploids, while to the frequent non-correlation of morphological and cytogenetical data could be added information from the "exact" science of chemistry, which was often correlated with neither. Indeed, the greatly increased involvement of chemists and mathematicians promising new taxonomic Jerusalems undoubtedly enhanced the inadequacy felt by many biosystematists who, after half a century of endeavour, had not solved the problems of "classical taxonomy".

It was, however, inevitable that the more heady expectations for biosystematics should fail to be realized. As Merxmüller (1970:144) wrote:

> what I still hardly understand is why it is exactly the biosystematists who press for a strongly generalised and fixed definition of that phase of evolution know as "species": they should be in the best position to appreciate the manifold variety of evolutionary processes and their widely inadequate investigation.

Taxonomy has always, amoeba-like, ingested those concepts and techniques which have proved useful and the increasing development of a broadly based "general purpose taxonomy", drawing on as many sources as possible (cf. Davis and Heywood, 1963), has permitted an appreciation of the important, but not overriding, role of bio-systematics and, indeed, chemotaxonomy and numerical taxonomy. Chromosome studies, central to much of biosystematics, have contributed significantly to our understanding of the evolution, differentiation and delimitation of populations and taxa but, whilst frequently assisting the taxonomic decisions which have to be taken, they do not dictate these decisions—any more than do chemical profiles and cluster analysis.

SOURCES OF CHROMOSOME DATA

Information on chromosomes is published in a wide variety of genetical, cytological, plant breeding and taxonomic journals throughout the world and, as in so many other branches of science, access to this information can be difficult, time-consuming and costly. Various efforts have been made to publish in certain journals (e.g. *Madroño*, *Watsonia*, *Taxon*) occasional data on chromosome

C

numbers which might otherwise be dissipated or not published. The lists of chromosome numbers edited by Löve for the International Organization of Plant Biosystematists since 1964 have been the most successful and 55 lists have been published to date (Löve, 1977). Various compilations have attempted to bring together information on chromosome numbers and those of Darlington and Janaki-Ammal (1945), Darlington and Wylie (1955), Chiarugi (1960), Fabbri (1963), Federov (1969) and Löve et al. (1977) have proved invaluable, while the annual *Index to Plant Chromosome Numbers* (Cave, 1958–1966; Ornduff, 1967–1969; Moore, 1970–1974) permits the necessary updating. However, the immense labour of preparing such compendia means that they must inevitably be uncritical and they tend to perpetuate cytological and taxonomic inaccuracies.

Much more careful editing, and subsequent correction of errors, is possible in regional surveys, such as those for Czechoslovakia (Májovský et al., 1976 and earlier), Iceland (Löve and Löve, 1956), Poland (Skalińska et al, 1971 and earlier) and Portugal (e.g. Barros Neves, 1973), or studies of particular groups, such as Compositae (Solbrig, 1977), Umbelliferae (e.g. Constance et al., 1976), *Astragalus* (Ledingham, 1960; Spellenberg, 1976) and *Anthurium* (Sheffer and Kamemoto, 1976). Data on chromosome numbers predominate in these compilations, and information on karyotype structure and meiotic behaviour is not summarized in any easily accessible form, although regional studies illustrated by idiograms and photographs—for example, those for New Zealand (e.g. Groves and Hair, 1971) and France (Gagnieu, 1967–1972)—provide pertinent data, as do many surveys of particular taxa, exemplified by Commelinaceae (Jones and Jopling, 1972), Proteaceae (Johnson and Briggs, 1963), *Narcissus* (e.g. Fernandes, 1966) and *Crocus* (e.g. Brighton, 1976a).

Electronic data processing can provide a continually updated bank of information on chromosome numbers, and some recent publications have demonstrated the feasibility of this (Moore, 1974; Löve and Löve, 1974; Solbrig, 1977). Despite considerable taxonomic and cytological editing of some of these lists, however, they show many of the disadvantages of the more traditional compilations already referred to, but the coming years will certainly see an increasingly sophisticated use of such data banks (Raven, 1975).

The outlook for other sorts of chromosomal data is not so clear. Access to the available information is limited not only, as noted above, because it is more widely dispersed but also because there is little general agreement as to how it can be most effectively presented. Information on chromosome numbers is so readily conveyed that it is now customary to include it in the stylized, truncated descriptions afforded species in Floras. On the other hand, the formulae utilized by, for example, Fernandes (1966) to convey relatively simple information on the number, size and centromeric position of chromosomes in the karyotype, have not been widely employed; nor have the devices for indicating patterns of heterochromatic and euchromatic segments in populations (e.g. Kurabayashi, 1957), a situation which should be faced in view of the increasing use of Giemsa and associated techniques in cytological studies relevant to taxonomic problems (Greilhuber and Speta, 1976). For these data, and for those on chromosome behaviour at meiosis in species and in both natural and artificial hybrids, there is no easy recourse at present except to scan the abstracting journals for indications under the taxa of special interest that such information is available. One of the major problems facing those interested in the chromosomes and plant taxonomy is to devise some acceptable system of summarizing these data so that not only can they be more readily made widely available but also incorporated as clearly as information on chromosome numbers into Floras and monographs.

EXTENT AND USE OF CHROMOSOME DATA

During the first quarter of this century chromosome data were relatively sparse (Fig. 1), so that their relevance to taxonomic problems could not be properly assessed. The markedly increased amount of information during the next 25 years demonstrated the widespread nature of basic cytogenetic mechanisms involved in the evolution and delimitation of taxa, while the spectacular rise in the publication of chromosome data during the past 25 years has begun to make clear the variety of ways in which these basic mechanisms are expressed in the great diversity of plants comprising the world's flora. What, then, is the relation between chromosomes and taxonomy today?

Extent of chromosome data

As hitherto, much the greatest amount of information is available on chromosome numbers, the most readily determined feature of the karyotype, and it is the easiest to assess, because of the well organized compendia and annual listings mentioned earlier. Well over 3000 chromosome counts have been published annually for the last 10 years or so (Fig. 1) and counts are now available for about 20% of species in the pteridophytes (Löve *et al.*, 1977) and, I estimate, about 15–20% of angiosperm species. There is a great regional imbalance in the level of information. As might be expected from the distribution of resources and taxonomic priorities, chromosome numbers are known for a much greater proportion of temperate than tropical species, the plants of tropical America being particularly in need of

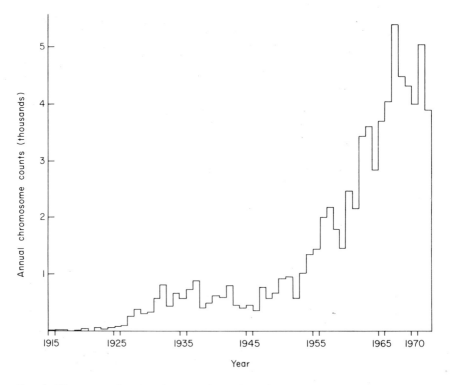

FIG. 1. Histogram showing the annual number of chromosome counts during the period 1915–1972, according to the standard lists (for references see text). Partly after Moore (1968).

cytological study (Raven, 1975), but there is still much work to be done in temperate floras. The floras of relatively small temperate regions have a reasonable primary coverage; for example, all species have at least one chromosome count in Iceland (Löve and Löve, 1956) and 42% in the British Isles (D. M. Cranston, personal communication)—but the proportion is less in larger areas and richer floras; counts are available for 36% of the indigenous species of New Zealand (Moore and Edgar, 1970), while moderately reliable chromosome counts are available for about 36% of the species (excluding monocotyledons) recognized in *Flora Europaea* (D. M. Moore, unpublished data), which deals with one of the best studied floras in the world.

In addition to regional variations in even primary chromosome data, there is considerable imbalance between different taxonomic groups. Raven (1975) lists 93 families, with some 141 genera, for which there is no cytological information whatsoever. Chromosome numbers are available for about 39% of the 20,000 species of Compositae (Solbrig, 1977), a family which has been the subject of intensive programmes of cytological study in the past two decades, for about 30% of the 840 species of the Umbelliferae (Moore, 1971), but only 5% (350/7000) of species in the Euphorbiaceae (Hans, 1973). The same sort of variation in knowledge is evident at the generic level, whilst it must be emphasized that only a relative handful of all these species are known from more than a single chromosome count. Studies on intraspecific variation in, for example, *Claytonia virginica* L. (Lewis, 1970), only serve to highlight this problem.

On turning to other sorts of chromosome data, the lack of any mechanism for summarizing information on, for example, karyotype structure and behaviour at meiosis in species and hybrids, precludes any exact overview of the current state of knowledge. At a rough guess, I would estimate that for at most 1% of angiosperm species is there published information on the size, centromeric position and occurrence of satellites in chromosomes of the karyotype, while the disposition of heterochromatic segments, revealed by cold treatment or Giemsa and related techniques, is very much more poorly known. So also is the meiotic behaviour of chromosomes in interpopulational and interspecific hybrids.

One further point which needs emphasizing yet again is the quality of the information available. During the first four decades of this century chromosomal reports were rarely related to particular voucher specimens in herbaria, so that the identity of the plants studied cannot be verified (Moore, 1968; Raven, 1975), while the interpretation of chromosome data based on sectioned material, prevalent before 1920 and not uncommon subsequently, presents particular problems (Raven, 1975). The classic case of allopolyploid evolution in *Spartina*, for example, was widely cited for over 30 years before Marchant (1963) showed the data, though not the general story, to be incorrect. Furthermore Merxmüller (1970) pointed out, that of the data summarized for *Myosotis* by Tischler (1950), on subsequent study 23·8% proved correct, 28·6% were cytologically incorrect, 23·8% were taxonomically incorrect and 23·8% were both taxonomically and cytologically incorrect. To what extent the approximately 121,000 chromosome counts now available for flowering plants and pteridophytes need revision only continuing adequately documented studies will reveal.

Use of chromosome data

The most widely accepted aims of modern taxonomy would seem to be: (1) to provide a convenient method of identification and communication; (2) to provide a classification which as far as possible expresses the natural relationships of organisms; and (3) to detect evolution at work, discovering its processes and interpreting its results (Davis and Heywood, 1963). The first two constitute the phenetic and the third the phylogenetic approach to taxonomy, and chromosome data play an important role in both. It is an oversimplification to try to differentiate consistently between the phyletic and phylogenetic information derived from the chromosomes, the carriers of the genetic material (e.g. Moore, 1968), but it is also too facile to give chromosome data an overriding importance in taxonomy, and claims that different chromosome numbers (n) indicate specific differences while different basic chromosome numbers (x) signal boundaries between genera (Löve and Löve, 1974) cannot be uniformly justified or applied.

> The chromosomes can determine many of the processes involved in evolutionary change or can act as markers of them, but they do not do so equally in all groups or in all situations (Moore, 1976:55).

Furthermore, the fullest cytogenetical information can only be obtained where hybridization and progeny analysis are possible, and these are most likely at the lower taxonomic levels. Consequently, there is a disparity in the potential taxonomic value of chromosome data in different groups and at different levels of the taxonomic hierarchy.

Despite the deficiencies already noted, the continued increase in chromosome data means that it is now possible to consider them seriously in relation to the taxonomy of major taxa. Families and taxa of higher rank can only be compared if the original basic chromosome numbers (x) of the taxa can be deduced and, since only absolute numbers, incapable of experimental analysis at these taxonomic levels, are available, there are obvious constraints (Raven, 1975). The original chromosome number of the angiosperms is considered be $x = 7$ (Grant, 1963; Stebbins, 1966; Raven et al, 1971). Among gymnosperms only $Ephedra$ $(x = 7)$ and $Welwitschia$ $(x = 21)$ seem chromosomally related to the angiosperms, but on other grounds it is doubtful if they share common ancestry (Raven, 1975) and the chromosomes provide little solution to the "abominable mystery". All major groups of angiosperms, except the subclass Caryophyllidae $(x = 9)$ of the Dicotyledoneae, seem to have $x = 7$, with $x = 11$–14 as particularly frequent derivatives, so that there is a "limited utility of chromosome numbers in question of the placement of particular families . . . only the most unusual chromosome number will tend either to confirm or deny a particular suggestion of affinity" (Raven, 1975:756). The various proposals for the ordinal grouping of families will consequently have to be judged on other than cytological criteria, although Raven's comprehensive survey does show their use in some instances. To take one example, the chromosomes support the inclusion of the Cyrillaceae and the Clethraceae in the Theales (Thorne, 1968) rather than in the Ericales (Cronquist, 1968).

The classification of many families has been aided or substantiated by information on chromosome number and morphology. The Gramineae provide many classical examples of the value of chromosome data in tribal delimitation. $Spartina$, for example, was for long placed in the tribe Chlorideae $(x = 10)$, although its chromosomes $(x = 7)$ were at variance. Marchant (1968) showed the genus to have in fact $x = 10$ small chromosomes and the dilemma was solved. The removal of $Paeonia$, with $x = 5$ and very large chromosomes, from the Ranunculaceae to the distantly related Paeoniaceae, has clarified the boundaries of the former family, which Gregory (1941) showed to be dominated by groups of genera with either large chromosomes ($x = 8$) or small chromosomes ($x = 7$). Detailed cytological studies of the Proteaceae (Johnson and Briggs, 1963), have shown that the relatively homogeneous subfamily Grevilleoideae has $n = 14$ small chromosomes (with $n = 10, 11, 15$ as derived conditions), the heterogeneous subfamily Proteoideae has $n = 7$ large chromosomes or $n = 14$ small chromosomes (and derived karyotypes of large or small chromosomes with $n = 14, 13, 12, 10$ and 5), and the morphologically and cytologically primitive genus *Placospermum* has large chromosomes and $n = 7$, distinct from both. This emphasizes the need for assessing all kinds of available information—morphological, anatomical, developmental and chemical—to make the most use of cytological data.

Intrafamilial groupings often show overlapping spectra of presumed primary basic chromosome numbers, which may suggest that some genera are incorrectly placed, but which more frequently indicate that identical numbers can evolve by very different routes that can only be interpreted by using all available data. In the Umbelliferae, for example, the subfamilies Apioideae ($x = 11$), Saniculoideae ($x = 8$) and Hydrocotyloideae ($x = 8, 10$) are modally different, but most of these presumed primary basic numbers occur in all three (Moore, 1971). Within the Hydrocotyloideae, tribes are similarly distinct, but not in the Apioideae. A comparable picture is shown by the Compositae (Solbrig, 1977): the Anthemideae, Arctotideae, Astereae and Lactuceae most commonly have $n = 9$; the Heliantheae $n = 7, 9, 17, 18$; the Vernonieae and Helenieae $n = 17$; the Senecioneae $n = 10, 20$; the Mutisieae $n = 12$; and so on. Likewise, in the Leguminosae, the tribes Podalyrieae ($x = (6), 7, 8, 9$), Bossiaeae ($x = 8, 9$), Liparieae ($x = 9$) and Crotalarieae ($x = (7), 8, 9$) are modally distinct, as are the Genisteae ($x = 8, (9), 11, 12, 13$), in which, however, $x = 12$ is much the commonest number and this, together with the prevalence of polyploidy, makes it quite different from the other tribes formerly grouped together in Genisteae *sensu lato* (Polhill, 1976).

Chromosomal diversity can, like its uniformity, be taxonomically unhelpful, though indicative of evolutionary processes. In the Meliaceae, for example, the subfamily Swietenioideae "is morphologically rather uniform and ... yet it is extremely heterogeneous karyologically and there is little or no support for its subdivision" (Styles and Vosa, 1971: 487). Most genera in the subfamily seem to be palaeopolyploids and this, together with such examples as the palaeotetraploid origin of the Rosaceae subfamily Rosoideae (Sax, 1931) and of the Genisteae (*see above*), suggest that, although progressive evolution of the angiosperms may have been largely at the diploid level (Stebbins, 1950), it is not helpful to dismiss polyploidy, occurring in 43% of flowering plants (Grant, 1963), as mere "evolutionary noise" (Wagner, 1970).

In studies of genera and species the chromosomes can still reveal the full range of cytogenetical possibilities for understanding the delimitation, affinities and evolution of taxa. The classical study of *Crepis* (Babcock, 1947), in which the clarification of its generic limits was first suggested by chromosome number and morphology, has not been diminished by subsequent work on many genera. For example, Rousi (1973) has shown in *Leontodon* that data on basic number, chromosome length, centromeric position and the occurrence of satellites provide strong support for one of the various taxonomic treatments available. In this the former genus *Thrincia* ($x = 4$) is included at sectional level, along with section *Asterothrix* ($x = 4, 7$), in subgenus *Apargia* of *Leontodon*, differentiated from subgenus *Leontodon* which, among other things, has $x = 6$ and 7 and chromosomes of different morphology. Cytotaxonomic studies have demonstrated the basic division between Old World species of *Astragalus*, including their close New World relatives ($x = 8$, euploidy common), and strictly New World lineages ($x = 11, 12, 13, 14, 15$, euploidy rare) (Spellenberg, 1976). In *Mentha*, which has small, structurally uniform chromosomes, 13 of the 25 species are known cytologically and chromosome numbers provide strong support for the subdivision into sections *Audibertia* ($x = 9$), *Pulegium* ($x = 10, ? 5$), *Preslia* ($x = 18$) and *Mentha* ($x = 12$); on the available data the basic number of section *Eriodontes* ($2n = 72, c. 144$) must be purely speculative (Harley and Brighton, 1977). Furthermore, in this genus, whose taxonomy has been bedevilled by widespread hybridization and vegetative propagation, chromosome numbers give clear evidence on the origins of many of the hybrids.

Detailed cytogenetical analysis of inter- and intraspecific hybrids has provided convincing evidence that species formerly placed in the

genera *Clarkia, Godetia, Eucharidium* and *Phaeostoma* be united into
Clarkia (Lewis and Lewis, 1955). Subsequent intensive study by
Lewis and his collaborators over two decades has only enhanced the
naturalness of this genus, in the evolution of which chromosomal
repatterning has played such an important role, while work on
Lopezia (Plitmann *et al.*, 1975) in the same family (Onagraceae)
emphasizes the taxonomic value of such studies. Extensive data on
chromosome behaviour in interspecific hybrids gave little clue to the
major genic barriers between, and sometimes within, species of
Spergularia (Ratter, 1976), but information on the behaviour of the
meiotic chromosomes of *Gayophytum* has illuminated the taxonomic
treatment of this difficult genus (Lewis and Szweykowski, 1964). Here,
the morphologically distinguishable diploids give rise to polyploids
which resemble their diploid progenitors and also hybridize with each
other to obscure the original differences. The recognition of such
"polyploid pillar complexes" (Ehrendorfer, 1959) is only possible by
studying the chromosomes and has done much to explain, although not
provide a formal solution for, the taxonomic complexity of groups
within *Achillea, Galium* and, unusually, *Perideridia*, in which the
diploid populations were most difficult to separate and were therefore
treated as a single polymorphic species, while the polyploid taxa were
reasonably distinct (Chuang and Constance, 1969).

The use of genome analysis for unravelling the relationships, if not
the taxonomic circumscription, of species within a genus has fallen
on hard times in recent years. In part, this is undoubtedly due to the
revelations on the genetic control of pairing, but it probably also
results from the fact that such analysis has only worked in a handful of
groups, such as the ferns (e.g. Vida, 1972) and *Viola* sect. *Viola*
(Moore and Harvey, 1961). Concomitantly, there has been a marked
reduction in the popular pursuit of the 1940s and 1950s of spotting
the parents of presumed allopolyploid species. This is partly due to
the factors just noted and partly to the accumulation of new
information. For example, on sound morphological and cytological
grounds, *Poa annua* L. ($2n = 28$) was deduced to be an allotetraploid
derived from *P. infirma* Kunth ($2n = 14$) and *P. supina* Schrader ($2n
= 14$) (Tutin, 1957). The discovery of diploid populations of *P.
annua* with regular bivalent formation and observations on somatic
karyotypes (Ellis *et al.*, 1970), has reopened the problem of the
origins of this species. Conversely, recent incisive studies, such as

those of Khyos (1965) in *Chaenactis*, provide clear chromosomal documentation for the origin of species.

It must be re-emphasized that the number, size, morphology and pairing relationships of chromosomes, individually or collectively of proven taxonomic value, are not equally of importance in all groups of plants (cf. Moore, 1968) and cytologically uniform genera, such as *Bobartia* (Strid, 1974), continue to show where the chromosomes are not particularly helpful taxonomically. However, the acquisition of new data can dramatically alter the situation. *Epilobium* sect. *Epilobium*, always with 18, small metacentric chromosomes in the somatic karyotype and virtually no evidence of chromosomal re-patterning, has for years been considered cytologically uninteresting. Recent extensive interspecific crosses (Seavey and Raven, 1977a, b) have shown that chromosomal arrangements, differing by one or two translocations, identify species from primarily Eurasia, Africa and Australasia (AA arrangement), North and South America (BB) and the circumboreal Alpinae (CC), thus providing a clear insight into the relationships and phytogeography of this apparently homo-geneous large group of species. Similarly, the species of *Anacyclus* have very similar karyotypes, but these have recently been shown (Schweizer and Ehrendorfer, 1976; Ehrendorfer *et al.*, 1977) to be distinguishable on Giemsa C-banding patterns. Furthermore, perennial species (sect. *Pyrethraria*) are different from the annuals (sect. *Anacyclus*), and within the latter two groups can be recognized. These sections and groups, recognizable by C-banding patterns, are largely confirmed by morphology, flavonoid profiles and inter-fertility, which also generally concur in indicating phylogenetic trends within the genus.

At the species level, all the chromosomal data referred to above and in the standard biosystematic literature, have proved to be of value for descriptive or interpretative taxonomy in some groups and not in others. Species, defined morphologically, usually appear to produce the genetical variability they normally require in populations against a background of stable chromosome numbers, although examples such as *Claytonia virginica* L., with 50 different cytotypes between $2n = 12$ and $2n = c.$ 191, provide a salutary warning (Lewis, 1970) that the chromosomes are not always so accommodating. Intrapopula-tional variation in the number, size and centromeric position of chromosomes in the karyotypes of, for example, *Crocus* (cf. Brighton,

1976b), in heterochromatic patterns revealed by cold treatment (cf. Kurabayashi, 1957) or C-banding, e.g. *Tulipa* (Filion, 1974), draw attention to the way in which the chromosomes mark the variation and evolutionary pathways within species, providing an insight into the evolution, but not the formal taxonomic treatment, of infraspecific entities.

That the continuing accumulation of data from the chromosomes adds to our taxonomic insight, while destroying previous *mores*, is well illustrated by a final example from *Clarkia*. "Seldom has a species been described with more knowledge than that on which *C. nitens* Lewis & Lewis was based" (Lewis and Bloom, 1972). It was distinct from *C. speciosa* Lewis & Lewis on numerous morphological differences and on the sterility of interspecific hybrids, which indicated chromosomal differences of at least seven interchanges. However, subsequent detailed studies revealed that morphological discontinuities marked one zone of contact between the species, cytological features marked another discontinuity 100 km distant, and populations in the intervening area suggested massive introgression, with selection for inter-fertility indicating that two species, formerly isolated, now share a common evolutionary future.

Despite a healthy scepticism of the value of the chromosomes in easing taxonomic decision-making, the annual outpouring of cytological data continues to illuminate the taxonomic study of many groups. With increasing knowledge the chromosomes will further reflect the manifold ways in which evolution has proceeded and will thus become an increasingly refined tool, among the many available to the discriminating taxonomist.

REFERENCES

Babcock, E. B. (1947). The genus *Crepis*. Pt. 1. The taxonomy, phylogeny distribution and evolution of *Crepis*. *Univ. Calif. Publs. Bot.* **21**, 1–197.

Barros Neves, J. (1973). Contribution à la connaissance cytotaxinimique des Spermatophyta du Portugal. VIII. Liliaceae. *Bolm Soc. broteriana*, Sér. 2, **47**, 157–212.

Brighton, C. A. (1976a). Cytology of *Crocus olivieri* and its allies. *Kew Bull.* **31**, 209–217.

Brighton, C. A. (1976b). Cytological problems in the genus *Crocus* (Iridaceae): I. *Crocus vernus* aggregate. *Kew Bull.* **31**, 33–46.

Camp, W. H. (1951). Biosystematy. *Brittonia* 7, 113–127.

Cave, M. S. ed. (1958–1966). "Index to Plant Chromosome Numbers 1956–64". University of North Carolina Press, Chapel Hill.

Chiarugi, A. (1960). Tavole chromosomiche della Pteridophyta. *Caryologia* 13, 27–150.

Chuang, T. I. and Constance, L. (1969). A systematic study of *Perideridia* (Umbelliferae–Apioideae). *Univ. Calif. Publs Bot.* 55, 1–74.

Constance, L., Chuang, T. I. and Bell, L. R. (1976). Chromosome numbers in Umbelliferae. V. *Am. J. Bot.* 63, 608–625.

Cronquist, A. (1968). "The Evolution and Classification of Flowering Plants." Nelson, London.

Darlington, C. D. and Janaki-Ammal, E. K. (1945). "Chromosome Atlas of Cultivated Plants." Allen and Unwin, London.

Darlington, C. D. and Wylie, A. P. (1955). "Chromosome Atlas of Flowering Plants." Allen and Unwin, London.

Davis, P. H. and Heywood, V. H. (1963). "Principles of Angiosperm Taxonomy." Oliver and Boyd, Edinburgh and London.

De Wet, J. M. J. (1968). Diploid–tetraploid–haploid cycles and the origin of variability in *Dichanthium* agamospecies. *Evolution* 22, 394–397.

De Wet, J. M. J. and Harlan, J. R. (1972). Chromosome pairing and phylogenetic affinities. *Taxon* 21, 67–70.

Ehrendorfer, F. (1959). Differentiation–hybridization cycles and polyploidy in *Achillea*. *Cold Spring Harb. Symp. quant. Biol.* 24, 141–152.

Ehrendorfer, F., Schweizer, D. Greger, H., and Humphries, C. J. (1977). Chromosome-banding and synthetic systematics in *Anacyclus* (Asteraceae-Anthemideae). *Taxon* 26, 387–394.

Ellis, W. M. Calder, D. M. and Lee, B. T. O. (1970). A diploid population of *Poa annua* L. from Australia. *Experientia* 26, 1156.

Fabbri, F. (1963). Primo supplemento alle Tavole chromosomiche della Pteridophyta di Alberto Chiarugi. *Caryologia* 16, 237–335.

Federov, A. A. (ed.) (1969). "Chromosome Numbers of Flowering Plants." Akademija Nauk SSSR, Leningrad.

Fernandes, A. (1966). Nouvelles études sur la section *Jonquilla* DC. du genre *Narcissus* L. *Bolm Soc. broteriana*, Sér. 2 40, 207–248.

Filion, W. G. (1974). Differential Giemsa staining in plants. I. Banding patterns in three cultivars of *Tulipa*. *Chromosoma (Berl.)* 49, 51–60.

Gagnieu, A. (ed.) (1967–1972). "Informations annuelles de caryosystématique et cytogénétique", Nos 1–6. Lab. Phytogénétique, Strasbourg and Lille.

Grant, V. (1963). "The Origin of Adaptations." Columbia University Press, London and New York.

Gregory, W. C. (1941). Phylogenetic and cytological studies in the Ranuculaceae Juss. *Trans. Am. Phil. Soc.*, N.S. 31, 443–520.

Greilhuber, J. and Speta, F. (1976). C-banded karyotypes in the *Scilla hohenackeri* group, *S. persica* and *Puschkinia* (Liliaceae). *Plant Syst. Evol.* 126, 149–188.

Groves, B. E. and Hair, J. B. (1971). Contribution to a chromosome atlas of the New Zealand flora—15. Miscellaneous families. *N.Z.J. Bot.* **9**, 569–575.

Hans, A. S. (1973). Chromosomal conspectus of the Euphorbiaceae. *Taxon* **22**, 591–636.

Harley, R. M. and Brighton, C. A. (1977). Chromosome numbers in the genus *Mentha* L. *Bot. J. Linn. Soc. (Lond.)* **74**, 71–96.

Huxley, J. S. (1940). "The New Systematics." Clarendon Press, Oxford.

Johnson. L. A. S. and Briggs, B. G. (1963). Evolution in the Proteaceae. *Austral. J. Bot.* **11**, 21–61.

Jones, K. (1970). Chromosomal changes in plant evolution. *Taxon* **19**, 172–179.

Jones, K. and Jopling, C. (1972). Chromosomes and classification of the Commelinaceae. *Bot. J. Linn. Soc. (Lond.)* **65**, 129–162.

Khasha, K. J. (ed.) (1974). "Haploids in Higher Plants." University of Guelph, Ontario.

Khyos, D. W. (1965). The independent aneuploid origin of two species of *Chaenactis* (Compositae) from a common ancestor. *Evolution* **19**, 26–43.

Kurabayashi, M. (1957). Evolution and variation in *Trillium*. IV. Chromosomal variation in natural populations of *Trillium kamtschaticum* Pall. *Jap. J. Bot.* **16**, 1–45.

Ledingham, G. F. (1960). Chromosome numbers in *Astragalus* and *Oxytropis*. *Can. J. Genet. Cytol.* **2**, 119–128.

Lewis, H. and Bloom, W. L. (1972a). The loss of species through breakdown of a chromosomal barrier. *Symp. Biol. Hung.* **12**, 61–64.

Lewis, H. and Bloom, W. L. (1972b). Interchanges and interpopulational gene exchange in *Clarkia speciosa*. *Chromosomes Today* **3**, 268–284.

Lewis, H. and Lewis, M. E. (1955). The genus *Clarkia*. *Univ. Calif. Publs Bot.* **20**, 241–392.

Lewis, H. and Szwekowski, J. (1964). The genus *Gayophytum* (Onagraceae). *Brittonia* **16**, 343–391.

Lewis, W. H. (1970). Extreme instability of chromosome number in *Claytonia virginica*. *Taxon* **19**, 180–182.

Löve, Á. (ed.) (1977). I.O.P.B. chromosome number reports LV. *Taxon* **26**, 107–109.

Löve, Á. and Löve, D. (1956). Cytotaxonomical conspectus of the Icelandic flora. *Acta Hort. Gotob.* **20**, 65–291.

Löve, Á. and Löve, D. (1974). "Cytotaxonomical Atlas of the Slovenian Flora." Cramer, Koenigstein.

Löve, A., Löve, D. and Pichi-Sermolli, R. E. G. (1977). "Cytotaxonomical Atlas of the Pteridophytes." Cramer, Koenigstein.

Májovský, J. (1976). Index of chromosome numbers of Slovakian flora (Part 5). *Acta F.R.N. Univ. Coment. (Bot.)* **25**, 1–18.

Marchant, C. J. (1963). Corrected chromosome numbers for *Spartina* x *townsendii* and its parent species. *Nature, Lond.* **199**, 929.

Marchant, C. J. (1968). Evolution in *Spartina* (Gramineae) III. Species chromosome numbers and their taxonomic significance. *Bot. J. Linn. Soc. (Lond.)* **60**, 411–417.

Merxmüller, H. (1970). Provocation or biosystematics. *Taxon* **19**, 140–145.

Moore, D. M. (1968). The karyotype in taxonomy. *In* "Modern Methods in Plant Taxonomy" (V. H. Heywood, ed.), pp. 61–75. Academic Press, London and New York.

Moore, D. M. (1971). Chromosome studies in the Umbelliferae. *In* "The Biology and Chemistry of the Umbelliferae" (V. H. Heywood, ed.), pp. 233–255. Academic Press, London and New York.

Moore, D. M. (1976). "Plant Cytogenetics." Chapman and Hall, London.

Moore, D. M. and Harvey, M. J. (1961). Cytogenetic relationships of *Viola lactea* Sm. and other West European arosulate violets. *New Phytol.* **60**, 85–95.

Moore, L. B. and Edgar, E. (1970). "Flora of New Zealand" Vol. II. Govt. Printer, Wellington.

Moore, R. J. (ed.) (1970–1974). Index to plant chromosome numbers for 1968–1972. *Reg. Veg.* **68, 77, 84, 90, 91**.

Orndff, R. (ed.) (1967–1969), Index to plant chromosome numbers for 1965–1967. *Reg. Veg.* **50, 55, 59**.

Plitman, U., Raven, P. H., Tai, W. and Breedlove, D. E. (1975). Cytological studies in Lopezieae (Onagraceae). *Bot. Gaz.* **136**, 322–332.

Polhill, R. M. (1976). Genisteae (Adans.) Benth and related tribes (Leguminosae). *Bot. System.* **1**, 143–368.

Ratter, J. A. (1976) Cytogenetic studies in *Spergularia*: IX. Summary and conclusions. *Notes Roy. bot. Gdn. Edinb.* **34**, 411–428.

Raven, P. H. (1974). Plant systematics 1947–1972. *Ann. Miss. bot. Gard.* **61**, 166–178.

Raven, P. H. (1975). The bases of angiosperm phylogeny: cytology. *Ann. Miss. bot. Gard.* **62**, 725–764.

Raven, P. H. and Thompson, H. J. (1964). Haploidy and angiosperm evolution. *Amer. Nat.* **98**, 251–252.

Raven, P. H., Kyhos, D. W. and Cave, M. S. (1971). Chromosome numbers and relationships in Annoniflorae. *Taxon* **20**, 479–483.

Riley, R. (1966). Genetics and the regulation of meiotic chromosome behaviour. *Sci. Progr. Lond.* **54**, 193–207.

Rousi, A. (1973). Studies on the cytotaxonomy and mode of reproduction of *Leontodon* (Compositae). *Ann. Bot. Fenn.* **10**, 201–215.

Sax, K. (1931). The origin and relationships of Pomoideae. *J. Arnold Arb.* **12**, 3–22.

Schweizer, D. and Ehrendorfer, F. (1976). Giemsa banded karyotypes, systematics and evolution in *Anacyclus* (*Asteraceae—Anthemideae*). *Plant Syst. Evol.* **126**, 107–148.

Seavey, S. R. and Raven, P. H. (1977a). Chromosomal evolution in *Epilobium* sect. *Epilobium* (Onagraceae). *Plant Syst. Evol.* **127**, 107–119.

Seavey, S. T. and Raven, P. H. (1977b). Chromosomal differentiation and the sources of the South American species of *Epilobium* (Onagraceae). *J. Biogeog.* **4**, 55–59.

Sheffer, R. D. and Kamemoto, H. (1976). Chromosome numbers in the genus *Anthurium*. *Am. J. Bot.* **63**, 74–81.

Skalińska, M., Jankun, A., and Weiso, H. (1971). Studies in chromosome numbers of Polish angiosperms. Eighth contribution. *Acta Biol. Cracov.* **14**, 55–102.

Solbrig, O. T. (1977). Chromosome cytology and evolution in the family Compositae. *In* "The Biology and Chemistry of the Compositae" (V. H. Heywood, J. B. Harborne and B. L. Turner, eds), pp. 246–260. Academic Press, London and New York.

Spellenberg, R. (1976). Chromosome numbers and their cytotaxonomic significance for N. American *Astragalus* (Fabaceae). *Taxon* **25**, 463–476.

Stebbins, G. L. (1950). "Variation and Evolution in Plants." Columbia University Press, London and New York.

Stebbins, G. L. (1966). Chromosomal variation and evolution. *Science, N.Y.* **152**, 1463–1469.

Strid, A. (1974). A taxonomic revision of *Bobartia* L. (Iridaceae). *Op. Bot.* **37**, 1–44.

Styles, B. T. and Vosa, C. G. (1971). Chromosome numbers in the Meliaceae. *Taxon* **20**, 485–499.

Thorne, R. F. (1968). Synopsis of a putatively phylogenetic classification of the flowering plants. *Aliso* **6**, 57–66.

Tischler, G. (1950). "Die Chromosomenzahlen der Gefässpflanzen Mitteleuropas." W. Junk, The Hague.

Tutin, T. G. (1957). A contribution to the experimental taxonomy of *Poa annua* L. *Watsonia* **4**, 1–10.

Vida, G. (1972). Cytotaxonomy and genome analysis of the European ferns. *Symp. Biol. Hung.* **12**, 51–60.

Wagner, W. H. (1970). Biosystematics and evolutionary noise. *Taxon* **19**, 146–151.

Chapter 4

Breeding Systems, Variation Patterns and Species Delimitation

C. A. STACE

INTRODUCTION

The taxonomist, in presenting his classification of plants, seeks to summarize, in a concise and practical form, the results of his discoveries on the patterns of variation and on the interrelationships of the variants of the plants he has been studying. This statement, which appears to me utterly uncontentious, embodies two separate concepts or aims—that the classification should both reflect the pattern of natural variation accurately, and present it in a usable form. In practice, however, these two aims are achieved to very varying degrees, and the level of success in one aim is not necessarily positively correlated with that in the other. Indeed, they are often negatively correlated; in other words, the more closely a classification expresses the relationships (however these might be interpreted) of the taxa, the less practical and utilitarian it might be.

Classifications which seek to express relationships in terms of phylogeny or cladistics appear to be those most in danger of becoming unrealistic in practical terms. This is because the characteristics which appear to indicate phylogenetic pathways are frequently cryptic or semi-cryptic (e.g. chromosome number, form and morphology; chemical compounds; microscopic anatomy) and, conversely, the most obvious features of a plant are often among the most strongly adaptive ones and hence among those most likely to have arisen in a parallel or convergent pattern in response to the changing environment. At the other extreme, wholly artificial classifications, of which Linnaeus' "sexual system" is a prime example, may be very easy to apply and excellent as pigeon-holing schemes, but usually they scarcely reflect overall interrelationships.

Modern classifications, where they cannot embody both aims, are therefore frequently compromises between these two extremes, and there is nowadays a less widespread belief than, say, 20 or 30 years ago that the ideal classification should of necessity be phylogenetic (cf. Heywood, 1966). More often classifications are admittedly natural or phenetic, and predictivity has become the most important criterion. There are, of course, still many who claim that classifications should aim to reflect phylogeny above all else (cf. Johnson, 1968) and few taxonomists would argue that phylogenetic data should not be incorporated into classifications whenever appropriate. Many would claim that the ultimate phenetic classification, presumably obtained by some perfected numerical technique, would in fact be phylogenetic also, since the character-states which have arisen in a parallel or convergent manner would be outweighed by the others. Whether this is in fact true is, of course, not known.

The breeding system of a plant cannot be ignored by taxonomists, whether or not they deliberately seek to incorporate data of this nature in their classifications, because the breeding system dictates as strongly as any factor the pattern of variation shown by plants. Hence, even taxonomists who, either by choice or of necessity, work solely with pressed plants in an herbarium inevitably (albeit sometimes unwittingly) consider their specimens not simply as individuals but also as members of populations which together comprise the total resources of the respective taxa. Were this not the case taxa could (and would) be defined by their *degree* of variation, rather than by the position of *discontinuities* in variation, which are in fact the signs which taxonomists seek. Thus, in an apomictic genus, two plants which are held to represent different species often exhibit far fewer differences than two plants of a single species in a sexual, outbreeding genus. Such facts amply illustrate that classifications are usually an expression of variation patterns rather than mere catalogues of variation.

This chapter is a survey of ways in which taxonomists treat taxa at or around the species level which, particularly as a result of different breeding systems, exhibit different patterns of variation. In some cases the survey is a statement of current practice but in others different systems which have been used are assessed, and I have not hesitated to give my own views when this seems appropriate.

THE SEXUAL, OUTBREEDING, NON-HYBRIDIZING SPECIES

The species is commonly defined on the basis of one or both of two sets of criteria; a discrete phenetic unit, separated from others by a discontinuity in variation (so-called morphological or phenetic species); and a discrete breeding unit, separated from others by a breeding barrier (so-called biological species). Of course, the first criterion was originally the only one, and it was not until the genecological studies of Turesson and others little over 50 years ago that serious attempts were made to apply the second criterion. These attempts came from the discovery that the morphological species very often does coincide with the biological species; the possibility that in cases of difficulty the limits of the breeding unit might decide the issue naturally presented itself. In fact this hope is not generally realized. Interbreeding data (the limits of the gene pool) are frequently of great taxonomic significance, and on occasions can tip the balance towards one decision or another, but often they are not very useful as taxonomic criteria, at worst serving to increase the problem or at best simply explaining why taxonomic problems exist.

The notion that the extent of the gene pool can define plant species limits is reinforced by man's more intimate knowledge of animals (particularly himself), in which the phenetic and genetic units are in the great majority of cases the same. Interbreeding species of animals are now known in several groups, particularly birds and fishes, but even in these classes the breeding unit is still considered to be of great significance and few zoologists would define a species without prominent reference to it.

Many botanists also seek to define species wholly or partially on cytogenetic criteria (cf. Runemark, 1961; Lehman, 1971) and there remains considerable argument concerning the extent to which such data (*vis-à-vis* phenetic data) should be utilized in species delimitation. V. H. Heywood has repeatedly argued strongly that morphological criteria alone should define species limits, and he has rejected any definition based on reproductive isolation, despite the fact that morphologically defined species "will represent different kinds of evolutionary situations and will be equivalent only by designation" (Heywood, 1963). This argument is based on the view that species must be visually recognizable to have any widespread practical application, and that in any case our information is far too incomplete to apply the cytogenetic criterion in all but a minute proportion of cases. Raven (1976b) has recently added his weight to the argument that species should not be redefined in cytogenetic terms.

At the other end of the scale is the taxonomist who believes cytogenetic data to be of paramount importance as the specific criterion. The strongest arguments for this view in recent years have been made by A. Löve, who claims that virtually all chromosome races are phenetically distinguishable and that, since they are intersterile, they should be recognized as distinct species. Moreover, the general acceptance of the biological species "would soon change the ancient art of classification into the modern science of critical taxonomy" (Löve, 1964). Such a view has led to the recognition or resurrection of a great many species which are based mainly on cytological features alone (or together with directly related features such as pollen or stoma size), and which are hence of limited practical value. Nevertheless, the great majority of species currently separated largely on cytogenetic evidence were earlier recognized as taxonomic entities (though often not as species) on morphological grounds (Löve, 1951).

Even if an advantageous cytogenetic redefinition were theoretically feasible, it would not be possible on practical grounds, as has been emphasized by many taxonomists. For instance, many taxa do not reproduce sexually, so the genetic criterion is not applicable, and in many cases even where it is, its application, or that of the cytological criterion, leads to the definition of units considerably different from those recognizable by morphological criteria. Moreover, cytogenetic data are totally lacking for the great majority of species, particularly where they are allopatric, thus rendering the cytogenetic criterion quite impracticable in most cases. This latter difficulty is even more true of chemical and numerical criteria.

Much of the argument as to whether or not two taxa should be recognized as separate species can be attributed to the fact that there is no absolute definition of a species, despite innumerable attempts to provide one. In other words, it is still not possible to give an objective statement of the degree of difference between two taxa which is necessary for them to qualify for specific rank, nor even to give the terms in which such differences should be stated (morphological, cytogenetic, mathematical etc.). This situation is most frustrating when taxonomists are attempting to communicate with non-taxonomists. All taxonomists must have met with the incredulity of chemists, physicists, physiologists or the like upon finding that the taxonomist cannot define precisely any of the units into which he classifies his organisms, and their wonderment at the situation which would arise if they were similarly unable to define an aldehyde, an electron or a tracheid. This is one of the reasons for the rather widespread belief that taxonomy is an art rather than a science. Certainly the competent taxonomist displays an artistic talent, an intuitive flair; but no more than, say, the biochemist designing his next experiment or interpreting the results of his last, and in my opinion the artistic element in taxonomy is greatly overemphasized.

The frustration alluded to above has been the impetus behind many attempts to provide an objective measure of specific status, by the use of either new characters (notably anatomical, cytological, genetic and chemical) or new techniques (e.g. scanning electron microscopy, DNA hybridization, gas chromatography, taxometrics). It is in this light that the redefinition of species mainly in cytological or genetical terms by cytogeneticists should be viewed, not as a take-over bid of the term species by quasi-taxonomists as it has sometimes been regarded. In any case, their attempts have not been very widely successful, and so far the prospects of species redefinitions on the basis of DNA pairing, base ratios or base sequences by biochemists, or of phenon levels by numerical taxonomists, among other fields, seem no better.

One is forced to the view that there is no lode-star or philosopher's stone by which an ideal species definition can be drawn up. In more particular terms, it is not logical to rely more heavily on cytogenetic than on morphological data in the mistaken belief that the former indicate some more important or fundamental aspect of organisms; neither is it wise to use morphological at the expense of cytogenetic evidence because the latter is less readily observable. Both must take their place in the spectrum of evidence, and even the most obscure pieces of information must be used whenever available. This has always been a central theme in numerical taxonomy, but it is no less vital a goal for all other taxonomists. But it is this open-ended species concept, based on variable sets of data, which has led to the subjectivity in taxonomy, illustrated by the cynical definition of a species as "a group of individuals sufficiently distinct from other groups to be considered by taxonomists to merit specific rank". This concept of a species can be compared with the judging of an international gymnastic or ice-dancing championship. The marks of individual judges vary with their relative emphasis on different aspects, with their assessment of each of these aspects, and to some extent with their national bias, but the variation in the marks lies in quite a narrow band around a mean. Thus, even where there are no means of absolute definition, the status is understood within very narrow limits.

Although the need to use all sorts of information in taxonomic judgement without *a priori* weighting has been emphasized, this is not to say that all sorts of data are of the same nature. If one considers to what extent chromosome number and petal number, for instance, are comparable pieces of taxonomic evidence, one encounters two separate considerations. Firstly, they are in one sense strictly comparable, as they are both numerical expressions of structural features which have arisen by natural selection. But, secondly, they are quite different, because plants with different chromosome numbers are very probably unable to interbreed and exchange genetic material. A careful distinction has to be made, therefore, between the use of chromosome number simply as a taxonomic character, when it should be given no more (or less) weighting than petal number, and its use as an indicator of a genetic pattern at the population level, and hence as a valuable clue to the likely morphological pattern, which in turn helps the taxonomist in his decisions.

It is in this sense that it was stated in the introduction that taxonomists cannot ignore the breeding system, whether or not they work with living plants and whether or not they consciously use biosystematic data in constructing their classifications. The breeding system, both historically and contemporarily, plays the leading part in determining the evolutionary pathway and, at any one time, the pattern of phenetic variation. It is a manifestation of the interaction between the internal genetic mechanisms of the plant and the external environmental conditions, and thus it is the basic determinant of the pattern of variation which confronts the taxonomist (cf. Ornduff, 1969). Variations in the breeding system are the prime causes of the variation in the phenetic pattern, and it is because this variation is wide that it is not possible to lay down an objective definition of a species.

In a great many groups of plants the species as defined phenetically closely coincides with that defined genetically, e.g. many Leguminosae (*Trifolium, Vicia, Lathyrus, Medicago* etc.), many Umbelliferae (*Peucedanum, Bupleurum, Seseli, Eryngium* etc.), *Sedum, Campanula* and *Allium*, although there are exceptions in most of these genera. These are all sexual taxa, often outbreeders, and there are effective isolating barriers between the species. Such genera rarely pose intrinsic taxonomic problems (although extrinsic problems often exist due to our ignorance through lack of detailed investigations), but our experience of them proves of much wider value, for they provide us with an approximate yardstick with which to judge other genera where the concomitance of morphological and cytogenetic limits is not so exact.

However the species has been defined, and in whatever terms, it has always approximated to that breadth of grouping which, in a non-hybridizing sexual outbreeder, is shown by the breeding unit (ecospecies, hologamodeme). Individual definitions often fall short of or exceed this "ideal" but they never lose sight of it. This represents a major unifying concept in taxonomy, and belies the cynical view that a species is whatever a taxonomist considers it to be. In the following two sections I shall develop the argument that species limits in genera which display widely different breeding systems (and hence may present difficult taxonomic problems) should be defined by reference to the "ideal" situation.

HYBRIDIZING SPECIES

If the accepted system of species delimitation incorporated the breeding unit as the major decisive factor, there would of course be no interspecific hybrids. That this has never been seriously proposed by any modern taxonomist can be attributed to two main considerations.

Firstly, the question of whether or not two taxa hybridize is not a simple matter. If they can hybridize they might frequently produce fertile hybrids, whereas equally they might very rarely form hybrids, which are always quite sterile. All intermediate situations exist. In addition there are innumerable cases of hybrids which have been synthesized artificially but which do not exist in the wild or, less extreme, which arise spontaneously in cultivation or in areas to which one or other parent has been introduced by man, yet do not occur in areas where both parents are native. Moreover, in many instances the extent of hybridization varies greatly from one geographical or ecological area to another, and the ability of two species to hybridize or the fertility of their hybrids may depend upon the strain or genotype of the parents involved. Clearly, a rather complex set of conditions would need to be stipulated if the genetic criterion of specific rank were to be rigorously adopted.

Secondly, the morphological divergence between taxa which interbreed is often very wide, so that their recognition as a single species would result in a remarkable degree of intraspecific polymorphism. In certain notorious genera, hybridization between the species is very widespread, and, in these, genetically defined species would encompass many morphological species—even a whole genus in some cases. Good examples of this are found in *Salix*, *Euphrasia*, *Epilobium*, *Rosa*, *Rumex*, *Potamogeton* and *Dactylorhiza* among British genera (Stace, 1975b).

In *Epilobium* and *Dactylorhiza*, virtually every parental species combination exists, while in *Salix* an artificial hybrid incorporating 13 different parental species has been obtained (Nilsson, 1954). There would clearly be no useful purpose achieved by amalgamating all the species of, say, *Epilobium* (12 in Britain, 23 in Europe, about 160 in the world) as a single species, and the same can be safely stated for many other parallel cases. The degree of morphological divergence of taxa is often not very closely correlated with either the extent of hybridization or the fertility or success of hybrids. Of the above genera, hybrids between species of *Salix*, of *Euphrasia*, of *Rosa* and of *Dactylorhiza* with the same chromosome numbers (and sometimes with different ones) are highly fertile, and in suitable habitats hybrid swarms occur. In the genus *Geum*, Gajewski (1957) found that very many interspecific hybrids could be synthesized, and that a good number exist in the wild even between morphologically extremely distinct taxa (such as the British *G. rivale* L. and *G. urbanum* L.), sometimes from different sections.

Conversely, there are many examples known of taxa which differ little or not at all phenetically but which are unable to interbreed. Some of these examples are intersterile races, which might have arisen by a mutation, of a single species. This topic is discussed in more detail in the next section.

This frequent lack of agreement between the phenetic and genetic species has been the impetus behind the provision of hierarchical systems of biosystematic categories alternative to the genus, species etc. of classical taxonomy, notably those of Turesson (1922) and Danser (1929). These systems, as well as the deme terminology of Gilmour and Gregor (1939) (which by contrivance can be converted to a hierarchical system) are compared by Stace (1975a).

It seems fairly clear that a major reason for the frequent hybridization between morphological species of higher plants is that the reproductive isolating mechanisms are often of the sort that serve to reduce the frequency of mating rather than to reduce the actual ability to mate. A wide range of isolating mechanisms is known in higher plants (Table I), and a discussion of them is presented by Stace (1975a). Of the 15 recognized in Table I the first seven are pre-pollination (external) mechanisms. Very often species are maintained as separate entities by a range of mechanisms rather than by just one, but whenever isolation relies on one or more pre-pollination

TABLE I. *Isolating mechanisms in flowering plants.*

Prezygotic	Postzygotic
Pre-pollination (external) mechanisms	10. Seed incompatibility
1. Geographical isolation	11. F_1 hybrid inviability
2. Ecological isolation	12. Non-fitness of F_1 hybrids
3. Seasonal isolation	13. F_1 hybrid sterility
4. Temporal isolation	14. F_2 hybrid inviability or sterility
5. Ethological isolation	15. Non-fitness of F_2 and backcross
6. Mechanical isolation	hybrids
7. Breeding behavioural isolation	
Post-pollination (internal) mechanisms	
8. Gametophytic isolation	
9. Gametic isolation	

mechanisms alone there is always a good chance that they will on occasion be overcome. Raven (1976b) believes "this situation [the frequent production of hybrids] to be the prevalent one among flowering plants as a whole".

In fact the degree of interspecific hybridization in flowering plants is to some extent a measure of the success, in terms of speciation, of pre-pollination isolating mechanisms. There is no reason to consider species which have arisen by external mechanisms any less important or fundamental than those which have arisen by internal ones, nor to treat them taxonomically any differently, so long as they maintain a reasonable degree of distinctness.

As has been hinted at above, "a reasonable degree of distinctness" can be interpreted in terms of related non-hybridizing species. Thus, if, in any given genus, two taxa which hybridize are about as phenetically different as the other non-hybridizing species, then they should also be treated as separate species, but, if they are less distinct morphologically, then a recognition at some infraspecific rank seems more appropriate. For that reason *Geum rivale* and *G. urbanum* are rightly treated as separate species, the more so because in some parts of Europe their external isolating mechanisms are still very effective (Briggs and Walters, 1969). In the genus *Silene* the fully interfertile taxa usually known as *S. maritima* With. and *S. vulgaris* (Moench) Garcke are probably better considered as a single species; in *Flora Europaea*, Chater and Walters (1964) treated them as two of eight subspecies of *S. vulgaris*. In the same genus, the two dioecious species, *S. alba* (Miller) E. H. L. Krause and *S. dioica* (L.) Clairv., frequently hybridize to give fully fertile hybrids and this, coupled with their great phenetic similarity, suggests that they also might well be treated as subspecies of one species; this has apparently never been done.

The argument used above for uniting *S. vulgaris* and *S. maritima*, and *S. alba* and *S. dioica*, is based on their close similarity compared with the differences between most other (non-hybridizing) species in the genus. But in a genus such as *Silene*, in which interspecific hybridization is in general rare, a second argument can be used, namely the evidently closer genetic relationship of such pairs of taxa as indicated by their ability to interbreed. It is my belief that in situations where the morphological divergence of taxa is not markedly variable, the breadth of the breeding unit becomes an important taxonomic character in its own right. It would be reasonable, for example, to settle the dispute concerning the ranks of *Calystegia sepium* (L.) R. Br. and *C. silvatica* (Kit.) Griseb., variously considered separate species or only subspecies of *C. sepium*, by applying the results of wide-scale breeding experiments in the genus. The above two taxa are as distinct as many in the genus, yet they are fully interfertile (Stace, 1961). Unfortunately, in this case, the necessary experiments have not been carried out. In the genus *Medicago*, where hybridization is rare, the two very distinct species *M. falcata* L. and *M. sativa* L. frequently hybridize to give rise to hybrid swarms and introgressants, some of which are used agriculturally. This situation has led to the two taxa being amalgamated as two subspecies of *M. sativa* by some taxonomists (e.g. Tutin, 1968).

Hence the taxonomist might find the discontinuities he seeks better expressed in either the phenetic or the genetic variation. A logical application of these two sorts of criteria would lead to a more rational classification at the specific level in a great many genera. In some cases, however, these criteria do not provide convenient discontinuities, or the two sorts might lead to different answers, so that taxonomic problems remain. In the Onagraceae there is disagreement as to whether or not the genus *Epilobium* should include the species often placed in the genus *Chamaenerion* (correctly known as *Chamerion*, cf. Holub, 1972). The species of *Chamerion* are distinct from those of *Epilobium sensu stricto* on a number of characters, but scarcely more so than several other groups within the latter, which has led Raven (1976a), an authority on the family, to amalgamate these two genera. On the other hand all the species within *Epilobium sensu stricto* are, so far as is known, capable of interbreeding, and hybrids are widespread in nature, but there are very strong breeding barriers between these species and species of *Chamerion*. This latter fact is strong evidence for retaining two distinct genera, as is done by other authorities.

Whereas the occurrence of interspecific hybrids may sometimes serve as evidence for the amalgamation of species, some hybrids are often treated as separate species. This is surely warranted whenever hybrids "have developed a distributional, morphological or genetical set of characteristics which is no longer strictly related to that of their parents" (Stace, 1975a). In the British flora good examples are

afforded by *Circaea* × *intermedia* Ehrh., *Equisetum* × *moorei* Newm.,
E. × *trachyodon* A. Braun, *Symphytum* × *uplandicum* Nyman,
Potamogeton × *zizii* Koch ex Roth, *P.* × *nitens* Weber, ×
Ammocalamagrostis baltica (Schrad.) P. Fourn., × *Festulolium
loliaceum* (Huds.) P. Fourn. and several hybrids in the genera *Mentha*
and *Salix*. These particular examples have been chosen because they
are extreme cases in a spectrum of situations—they are relatively
simple hybrids, mostly with an obvious parentage and many probably straightforward F_1s, and many taxonomists would question the
wisdom of treating them like species. Less extreme is a great range of
other hybridogenous taxa about which there is more agreement for
their specific treatment. Among these are distinctive F_2 segregants
and introgressed variants in various genera; crypthybrids (ancient
crosses whose hybrid origin is only apparent from meiotic studies) in
Rosa (Blackburn and Harrison, 1924); translocation heterozygotes in
Oenothera (Cleland, 1972); diploid hybrids originating from
diploid–tetraploid crosses in *Euphrasia* (Yeo, 1956); and
innumerable agamospecies and amphidiploids. Even amphidiploids,
well known as special cases, are by no means always without their
problems. Not all of them are the stable new species we once believed,
but are segmental allopolyploids, polyphyletic assemblages of
morphologically similar individuals at one ploidy level, or rather
unstable and possibly short-lived polyploid segregates.

Many examples of the above sort are the source of taxonomic
discussion and argument. It is suggested, again, that decisions as to
their specific status or otherwise are best reached by a consideration
of whether or not the hybridogenous taxa have developed a distinctive morphological, genetic or distributional facies, and by reference
to related non-hybridogenous species.

Before leaving the subject of hybridization the taxonomic naming
of hybrids should be mentioned. The *International Code of Botanical
Nomenclature* allows for the naming of hybrids by a hybrid formula,
such as (to use an example from above) *Calystegia sepium* × *C.
silvatica* and/or by a binary name with a multiplication sign between
the generic and specific names, in the above case *C.* × *lucana*
(Tenore) G. Don. The use of the latter is allowed for "whenever it
seems useful or necessary"; this is, in the opinion of a certain
proportion of taxonomists, in every case. One special sort of hybrid,
"amphidiploids and similar polyploids", when treated as species,

may bear the binary name without the multiplication sign. In fact the *International Code* effectively also sanctions the use of a binary name without the multiplication sign for any other sort of hybrid which is treated as a species, because it does not deny the fact that species can have a hybrid origin. For hybrid taxa treated as species there are, therefore, three different methods of nomenclatural treatment. The arguments for and against the provision of a binary name for hybrid taxa are never ending and will probably never be resolved, but binary names are clearly "useful or necessary", to use the words of the *International Code*, whenever one wishes to treat the hybrid as a species, and also for those many cases where hybridity is certain but the parentage is not. The argument forwarded here is that rather more hybrids than is customary should be treated as species.

SEMI-CRYPTIC SPECIES

Under this heading I am including several unrelated categories of taxa which are generally recognized as species, but which resemble each other closely and are often difficult to distinguish, particularly when anything but a full set of evidence is available. The reasons for them being given specific rank, despite their phenetic closeness, are very diverse, but for the purposes of this discussion they may be considered together; all of them represent situations where the biological species is narrower than the morphological species, so they are in this sense the opposite of the cases discussed in the last section. Three major categories can be delimited:

1. Sexual taxa which are intersterile yet morphologically very close. The commonest reason for their intersterility is differing chromosome numbers, often coupled with external isolating mechanisms, such as different habitat preferences or geographical distributions, but there are examples, e.g. *Anagallis arvensis* L. and *A. foemina* Mill., where this is not the case. Whether or not one agrees with the dictum of Löve that virtually all "chromosome races" should be treated as separate species, a great many of such taxa are now generally recognized as separate species, whereas before their cytological background was known they were not. *Polypodium vulgare* L., *Nasturtium officinale* R. Br., *Empetrum nigrum* L., *Monotropa hypopitys* L., *Veronica hederifolia* L., *Galium palustre* L., *Juncus*

bufonius L. and *Poa pratensis* L. are good examples in the British
flora of taxa which contain plants of different chromosome number
(mostly in simple multiples), each of which are now commonly
considered separate species, although in almost all cases there are
some taxonomists who would regard the cytotypes as different
infraspecifically rather than specifically. In addition there are further
species containing different cytological races which are more usually
or always regarded only as infraspecific variants, e.g. *Ranunculus
ficaria* L., *Hypericum maculatum* Crantz, *Luzula multiflora* (Retz.)
Lejeune and *Eleocharis palustris* (L.) Roem. & Schult. It would
presumably be argued that the phenetic differences between the
cytotypes in the first group are greater than between those in the
second, which justifies the different taxonomic treatment, but the
fundamental situation is the same in both and it is extremely difficult
to draw a line between the two groups. In my opinion all these
examples should be treated similarly. Where the cytological races are
truly cryptic, however, they do not merit taxonomic recognition at
all, although Löve (1951) claimed that such situations are very rare.
More often the different chromosome races do show morphological
differentiation, but it is of a variable, quantitative type so that sharp
boundaries are not drawn. This is most common in cases of polyploid
complexes, e.g. *Cardamine pratensis* L., *Valeriana officinalis* L. and
Caltha palustris L., which may defy satisfactory taxonomic
delimitation.

2. Sexual taxa which are usually interfertile yet which rather rarely
hybridize in the wild due to their possession of an inbreeding system.
Plants which are largely or entirely inbreeding are apt to form local,
relatively homogeneous races which can often be recognized readily
by specialists as distinct taxa and have often been considered species
in their own right. In most cases the earliest taxonomists recognized
in each case only one species, but subsequent analysis showed that
each of these could be broken down into two or more (sometimes
hundreds of) constituent members. Species of this sort are *Capsella
bursa-pastoris* (L.) Medic., *Erophila verna* (L.) Chevall., *Arenaria
serpyllifolia* L., *Montia fontana* L., *Salicornia europaea* L., *Vicia
sativa* L., *Polygonum aviculare* L., *Senecio vulgaris* L. and
Zannichellia palustris L. A pioneer in the detailed study of inbreeding
species was A. Jordan (1814–1897), whose name is particularly asso-

ciated with the genus *Erophila*, in which he was able to recognize a great many separate taxa. These species of a lower order are commonly known as Jordanons, to distinguish them from the wider species or Linneons.

There are rather few species for which precise figures are available to indicate the degree of inbreeding as opposed to outbreeding. For *Senecio vulgaris* figures of 1% outbreeding have been calculated (Hull, 1974), while in *Vulpia microstachys* (Nutt.) Benth., Kannenberg and Allard (1967) concluded it was less than 0·01%. Even in the latter species considerable variation is detectable within the progeny of one individual, indicating either that a very low level of outbreeding can produce this variation or that the manifestation of a formerly outbreeding system can survive for a very long time. For these reasons Jordanons are nowadays almost universally not recognized taxonomically at any level. In my opinion they should be, especially as this would distinguish such Linneons from those in which inbreeding is equally characteristic but in which, at least at the morphological level, rather little differentiation has arisen, e.g. *Arabidopsis thaliana* (L.) Heynh., *Stellaria media* L. and *Valerianella locusta* (L.) Betcke.

3. Non-sexual taxa which reproduce apomictically, either by agamospermy or by vegetative apomixis. Certain agamospermous genera or infrageneric groups are notoriously difficult taxonomically, being represented by apparently innumerable pure lines which are propagated exactly by their peculiar, uniparental mode of reproduction. Perhaps the best-known European examples are in the genera *Hieracium*, *Taraxacum*, *Rubus*, *Alchemilla*, *Sorbus* and *Ranunculus*, but there are many more besides. As with the inbreeding Jordanons, it is possible for specialists to become acquainted with these agamospecies, and to unfailingly recognize them. They have therefore been given names, and the current practice is to recognize them all as distinct species ("microspecies") and provide them with binomials. Such a treatment is justified on the grounds that they remain more constant then the autogamous (self-pollinating) Jordanons, although this is true only for obligate apomicts. In some of these groups, however, agamospermy is facultative; sometimes species can reproduce sexually as well as agamospermously, e.g. *Hieracium umbellatum* L., whereas in other cases a few sexual species

can act as female parents in crosses with agamospermous ones, e.g. *Rubus ulmifolius* Schott. Where facultative apomixis is at all widespread the recognition of separate agamospecies is no easier than that of separate inbreeding Jordanons, and the taxa delimited are no more meaningful.

Vegetative apomixis less frequently leads to very complex taxonomic situations, but there are good examples where clones have become reasonably widespread and therefore repeatedly recognizable. As is also the case with agamospecies, such plants are usually sexually sterile as a result of past hybridization, and often have irregular cytological features. The genus *Mentha* is a good example. In others, however, vegetative propagation has become the predominant mode of reproduction because of unsuitable environmental factors, or because only one sex of a dioecious species is present, or because the sexual life cycle is long and sexual generations are thus produced much less frequently than vegetative progeny. The complexities of the genus *Ulmus* almost certainly result from a history of hybridization followed by clonal propagation of nothomorphs (new hybrid forms); Jeffers and Richens (1970) found 338 of their lowest-order groups among their 1131 samples from England, and 216 among 655 samples from northern France (Richens and Jeffers, 1975). These groups can, however, be reduced in number by successive clustering, and it should be possible to reach a stage where the clusters correspond with those recognizable by field botanists; such taxa would be as deserving of binomials as agamospecies.

In each of these three sorts of situations one can recognize two levels of taxon, both of which have at some time been considered binomial species. It is undeniable that in each case the wider taxon, or Linneon, is the closer equivalent of the "standard", outbreeding, sexual species, and if one wishes to retain some degree of equivalence between species in groups with different breeding systems it is that level, and not the lower one, which should be accorded specific rank. Such a treatment in the past has been associated with a "lumping" rather than a "splitting" philosophy, but this label is neither desirable nor justified, for the taxa at the lower level (Jordanons or microspecies) can and should be accorded taxonomic recognition.

If one scours the taxonomic literature one finds a multiplicity of schemes drawn up to satisfy the need to recognize taxa at the above two levels. The purpose of this chapter is not to review these methods, for it would take many pages, nor to add to them, but some examples are given in Table II and the need to adopt a universally acceptable system is stressed below.

TABLE II. *Examples of methods used to designate major and minor taxa at around the species level.*

Major taxon	Minor taxon	Reference
1. *Ranunculus* Cycle *Auricomi* Ovcz.	*R. auricomus* L.	Ovczinnikov (1937)
2. *Ranunculus ficaria* L.	*R. ficaria* var. *ficaria*	Marsden-Jones (1935)
3. *Ranunculus ficaria* L.	*R. ficaria* subsp. *ficaria*	Lawalrée (1955)
4. *Nigella arvensis* complex	*N. arvensis* L.	Strid (1970)
5. *Silene mollissima* group	*S. mollissima* (L.) Pers.	Chater and Walters (1964)
6. *Silene cucubalus* Wibel	*S. maritima* With. semi-species *cucubalus* Wibel	Valentine and Löve (1958)
7. Gestammtart (collective species) *Vicia sativa*	*V. sativa* L.	Ascherson and Graebner (1909)
8. *Monotropa hypopitys* L. agg.	*M. hypopitys* L.	Warburg (1952)
9. *Monotropa hypopitys* L. *sensu lato*	*M. hypopitys* L.	Dandy (1958)
10. *Hieracium* section *Umbellata*	*H. umbellatum* L.	Pugsley (1948)
11. *Juncus* series *Bufonii*	*J. bufonius* L.	Krechetovich and Goncharov (1935)

CONCLUSIONS

The foregoing sections have successively discussed species recognition in plants where the morphological and genetical species limits

D

coincide, where the genetical limits are wider than the morphological, and where the morphological limits are wider than the genetical. Taxonomic problems are few in the first situation, but common in the second two. In fact most intrinsic taxonomic problems in plants arise from situations of these kinds, although there are other origins as well; among these ought to be mentioned the translocation heterozygotes in the *Oenothera biennis* L. group (Cleland, 1972) and the unbalanced polyploids in the *Rosa canina* L. group (Blackburn and Harrison, 1924). These two groups are of the type called by Grant (1953) heterogamic complexes, a useful summary of which is given in his textbook (Grant, 1971). In addition there are problems created by clinal variation (continuous variation along a geographical or ecological gradient) (cf. Bocher, 1967), by the special patterns of variation found in some cultivated plants (cf. Harlan and de Wet, 1971); and by special combinations of circumstances which have led various taxonomists to claim that some genera do not lend themselves to orthodox taxonomic treatment, e.g. *Euphrasia* (see Chapter 8), *Quercus* (see Van Valen, 1976) and *Ulmus* (see Richens, 1955). Nevertheless, these are exceptional, and it is important to realize that the great majority of intrinsic taxonomic difficulties are, so far as we know, of the sort covered in the previous two sections.

With such a huge range of variation patterns to classify, it is not surprising that there is an enormous list of taxonomic ranks at and below the species level which have been proposed and used over the past 50 years; indeed, there seems to be no up-to-date catalogue of them all. The one presented by Sylvester-Bradley (1952) (thought to have been largely compiled by A. J. Wilmott) contained about 100 ranks, and was even then incomplete. A more complete list of ranks for cultivated taxa alone (over 50) was provided by Jirásek (1961). Some workers have attempted to bring about order by suggesting the use of a standardized set of concisely defined ranks, e.g. Du Rietz (1930), Wilmott (1949), Jirásek (1966), Jeffrey (1968), Harlan and de Wet (1971) and Sundermann (1975). Almost all of these attempts, with one exception (that of Du Rietz), are remarkable for their lack of impact on taxonomic practice, and for the non-appearance in the literature of most of the new ranks which were variously introduced. Again, this is hardly surprising in view of the great disparities between many of the recommended schemes.

The *International Code of Botanical Nomenclature* primarily recognizes species, subspecies, varieties, subvarieties, forms and subforms. In "sexual, outbreeding, non-hybridizing species" the use of the infraspecific terms subspecies, variety and form with the connotations given them by Du Rietz (1930) is in most cases adequate, and there has been a reasonable measure of agreement by taxonomists in this matter, but in groups with different sorts of variation patterns from these the provisions of the *International Code* are quite inadequate at around the species level as well as below it (Stace, 1976). One of the most unfortunate consequences of this inadequacy, coupled with the deeply entrenched adherence to the Du Rietz categories, is that the subspecies as conceived by Du Rietz (i.e. the geographical race) has become debased by its use for many different situations, in particular for groups of inbreeding taxa, groups of cultivars, and semi-cryptic cytotypes. Thus, in many Floras, the rank of subspecies now denotes nothing more precise than a major unit below the species level.

The dilemma in which taxonomists thus find themselves is one practical manifestation of the crisis in taxonomy described by Heywood (1973), and its solution is an urgent priority (cf. Raven, 1976b). The answer is clearly not to create many new ranks, for these would either be neglected or add to the confusion; a system too confusing to be applied or understood by a wide range of users has lost its status as a general purpose classification. Yet, equally clearly, some modifications to or re-drafting of the *International Code* are required. Such changes must pay attention to the need to reflect natural variation, to be readily applicable, to be easily understandable, to be flexible but unambiguous, and to be acceptable to the great majority of taxonomists (cf. Stace, 1976).

A system which catered adequately for the meaningful naming of semi-cryptic species would go a long way towards easing the situation, since plants showing this sort of taxonomic problem are so common. There are three main methods of approach: the use of the aggregate/segregate system advocated by Heywood (1963); the use of a new major infraspecific rank such as the semi-species favoured by Valentine and Löve (1958); or the invention of a new two-tier binomial terminology. It is clear that the adoption of a single agreed scheme is vital, but only if it is framed by an internationally recognized body will it gain universal acceptance.

REFERENCES

Ascherson, P. and Graebner, P. (1909). "Synopsis der Mitteleleuropäischen Flora", Vol. 6, pp. 959–975. W. Engelmann, Leipzig.

Blackburn, K. B. and Harrison, J. W. H. (1924). Genetical and cytological studies in hybrid roses, 1. *Br. J. exp. Biol.* **1**, 557–570.

Böcher, T. W. (1967). Continuous variation and taxonomy. *Taxon* **16**, 255–258.

Briggs, D. and Walters, S. M. (1969). "Plant Variation and Evolution." Weidenfeld and Nicolson, London.

Chater, A. O. and Walters, S. M. (1964). *Silene* L. *In* "Flora Europaea" (T. G. Tutin *et al.*, eds), Vol. 1, pp. 158–181. Cambridge University Press, Cambridge.

Cleland, R. E. (1972). "Oenothera: Cytogenetics and Evolution". Academic Press, London and New York.

Dandy, J. E. (1958). "List of British Vascular Plants". Botanical Society of the British Isles, London.

Danser, B. H. (1929). Über die Begriffe Komparium, Kommiskuum und über die Entstehungsweise der Konvivien. *Genetica* **11**, 399–450.

Du Rietz, G. E. (1930). The fundamental units of biological taxonomy. *Svensk bot. Tidskr.* **24**, 333–428.

Gajewski, W. (1957). A cytogenetic study of the genus *Geum* L. *Monographiae bot.* **4**.

Gilmour, J. S. L. and Gregor, J. W. (1939). Demes: a suggested new terminology. *Nature, Lond.* **144**, 333.

Grant, V. (1953). The role of hybridisation in the evolution of the leafy stemmed Gilias. *Evolution, Lancaster, Pa.* **7**, 51–64.

Grant, V. (1971). "Plant Speciation". Columbia University Press, New York and London.

Harlan, J. R. and de Wet, J. M. J. (1971). Toward a rational classification of cultivated plants. *Taxon* **20**, 509–517.

Heywood, V. H. (1963). The "species aggregate" in theory and practice. *Regnum veg.* **27**, 26–37.

Heywood, V. H. (1966). How many taxonomies? *Review roum Biol., Bot.* **11**, 101–106.

Heywood, V. H. (1973). Taxonomy in crisis? *Acta bot. Hung.* **19**, 139–146.

Holub, J. (1972). Taxonomic and nomenclatural remarks on *Chamaenerion* auct. *Folia geobot. phytotax. Bohemoslov.* **7**, 81–90.

Hull, P. (1974). Self-fertilization and the distribution of the radiate form of *Senecio vulgaris* L. in Central Scotland. *Watsonia* **10**, 69–75.

Jeffers, J. N. R. and Richens, R. H. (1970). Multivariate analysis of the English elm population. *Silvae Genet.* **19**, 31–38.

Jeffrey, C. (1968). Systematic categories for cultivated plants. *Taxon* **17**, 109–114.

Jirásek, V. (1961). Evolution of the proposals of taxonomical categories for the classification of cultivated plants. *Taxon* **10**, 34–45.

Jirásek, V. (1966). Systematika Kulturních rostlin a její třídicí kategorie. *Preslia* **38**, 267–284.

Johnson, L. A. S. (1968). Rainbow's end: the quest for an optimal taxonomy. *Proc. Linn. Soc. N.S.W.* **93**, 8–45.

Kannenberg, L. W. and Allard, R. W. (1967). Population studies in predominantly self-pollinated species, 8. Genetic variability in the *Festuca microstachys* complex. *Evolution, Lancaster, Pa.* **21**, 227–240.

Krechetovich, V. I. and Goncharov, N. F. (1935). *Juncus* L. *In* "Flora U.R.S.S." (V. L. Komarov, ed.), Vol. 3, pp. 504–559. Academy of Sciences of U.S.S.R., Leningrad.

Lawalrée, A. (1955). *Ranunculus* L. *In* "Flore Générale de Belgique" Vol. 2, pp. 55–103. Ministère de l'Agriculture, Bruxelles.

Lehman, H. (1971). Classification and explanation in biology. *Taxon* **20**, 257–268.

Löve, A. (1951). Taxonomical evaluation of polyploids. *Caryologia* **3**, 263–284.

Löve, A. (1964). The biological species concept and its evolutionary structure. *Taxon* **13**, 33–45.

Marsden-Jones, E. M. (1935). *Ranunculus Ficaria* Linn.: Life cycle and pollination. *J. Linn. Soc. Lond., Bot.* **50**, 39–55.

Nilsson, N. H. (1954). Über Hochkomplexe Bastardverbindungen in der Gattung *Salix*. *Hereditas* **40**, 517–522.

Ornduff, R. (1969). Reproductive biology in relation to systematics. *Taxon* **18**, 121–244.

Ovczinnikov, P. N. (1937). *Ranunculus* L. *In* "Flora U.R.S.S." (V. L. Komarov, ed.), Vol. 7, pp. 271–509. Academy of Sciences of U.S.S.R., Leningrad.

Pugsley, H. W. (1948). A prodromus of the British *Hieracia*. *J. Linn. Soc. Lond., Bot.* **54**, 1–356.

Raven, P. H. (1976a). Generic and sectional delimitation in Onagraceae, tribe Epilobieae. *Ann. Mo. Bot. Gdn* **63**, 326–340.

Raven, P. H. (1976b). Systematics and plant population biology. *Syst. Bot.* **1**, 284–316.

Richens, R. H. (1955). Studies on *Ulmus*, 1. The range of variation of East Anglian elms. *Watsonia* **3**, 138–153.

Richens, R. H. and Jeffers J. N. R. (1975). Multivariate analysis of the elms of northern France, 1. Variation within France. *Silvae Genet.* **24**, 141–150.

Runemark, H. (1961). The species and subspecies concepts in sexual flowering plants. *Bot. Notiser* **114**, 22–32.

Stace, C. A. (1961). Some studies in *Calystegia*. Compatibility and hybridisation in *C. sepium* and *C. silvatica*. *Watsonia* **5**, 88–105.

Stace, C. A. (1975a). "Hybridization and the Flora of the British Isles". Academic Press, London and New York.

Stace, C. A. (1975b). Wild hybrids in the British flora. *In* "European Floristic and Taxonomic Studies" (S. M. Walters and C. J. King, eds), pp. 111–125. Botanical Society of the British Isles, Faringdon.

Stace, C. A. (1976). The study of infraspecific variation. *Curr. Adv. Pl. Sci.* **8**, 513–523.

Strid, A. (1970). Studies in the Aegean flora, 16. Biosystematics of the *Nigella arvensis* complex. *Op. bot. Soc. bot. Lund.* **28**, 1–169.

Sundermann, H. (1975). Zum problem der Definition taxonomischer kategorien (Spezies, Subspezies, Praespezies, Varietät). *Taxon* **24**, 615–627.

Sylvester-Bradley, P. C. (1952). The classification and co-ordination of infra-specific categories. *Syst. Assocn, Taxonomic Principles Committee* **1/52.**

Turesson, G. (1922). The genotypical response of the plant species to the habitat. *Hereditas* **3**, 211–350.

Tutin, T. G. (1968). *Medicago* L. *In* "Flora Europaea" (T. G. Tutin *et al.*, eds), Vol. 2, pp. 153–157. Cambridge University Press, Cambridge.

Valentine, D. H. and Löve, A. (1958). Taxonomic and biosystematic categories. *Brittonia* **10**, 153–166.

Van Valen, L. (1976). Ecological species, multispecies, and oaks. *Taxon* **25**, 233–239.

Warburg, E. F. (1952). *Monotropa* L. *In* "Flora of the British Isles" (A. R. Clapham *et al.*), pp. 790–791. Cambridge University Press, Cambridge.

Wilmott, A. J. (1949). Intraspecific categories of variation. *In* "British Flowering Plants and Modern Systematic Methods" (A. J. Wilmott, ed.), pp. 28–45. Botanical Society of the British Isles, London.

Yeo, P. F. (1956). Hybridisation between diploid and tetraploid species of *Euphrasia*. *Watsonia* **3**, 253–269.

Chapter 5

Information Content of Keys for Identification

P. H. A. SNEATH and A. O. CHATER

INTRODUCTION

In the mid 1960s Professor Tutin often discussed with us the very varied amounts of information available on different groups of European flowering plants. He was especially conscious of the gaps in the available information on the genera of the Umbelliferae, a family he was then editing for *Flora Europaea* (Tutin *et al.*, 1964). In response to suggestions that it might lead to some profitable insight into the family, perhaps help in deciding debatable relationships between certain genera, and certainly help in producing a key to the genera, he prepared a table of the main discriminatory characters of the 110 European genera. This involved him in many weeks of arduous work with specimens as well as literature, for often a character recorded in the literature for one genus was ignored for the next, and adequate specimens showing, for example, fruits or characters of the roots, were often hard to find. It was not perhaps so widely appreciated then as it is now (Pankhurst, 1975) how very incomplete and lacking in comparative data even quite long and ostensibly full descriptions often were.

This exercise gave rise to a small numerical taxonomic study, which, while of interest to those working up the family for *Flora Europaea*, was not sufficiently detailed to be worth publishing. More importantly for us, however, the table of characters served as a basis for trying out some proposals for the automatic production of diagnostic keys, and several keys were made from it by the computer method of Pankhurst (1970). These keys turned out to be usable, although they did not work as well as Tutin's own key produced by orthodox methods. Since the actual writing of a key by orthodox methods is usually a rapid business once the necessary data are assembled, it is difficult to say whether the labour needed to produce the data for a computer-generated key is commensurate with the quality of the key thus produced. One could better answer this question if one knew the minimum number of characters needed to be scored in the table to permit the computer to construct an adequate key. We now offer some pointers to a solution of this problem.

In more general terms, this work in the Umbelliferae led us to ask a related question: how much information is needed to produce a workable identification system? We felt that any generalizations we could make would be valuable to those planning future taxonomic and diagnostic work. Some of the resulting ideas have already been of considerable significance in bacteriology, where the cost of diagnostic tests is high and where every effort must be made to keep them to a minimum (Sneath, 1969). The question of whether an economical key is worth aiming for in other groups is more debatable, and a program of key-checking and key-testing would be needed to weigh the advantages of brevity and neatness against those of practicality and reliability. We do, however, envisage that as more data are expressed in quantitative form for numerical methods in taxonomy (p. 385 in Sneath and Sokal, 1973), these questions will attract greater attention. Some theoretical work has been done on the minimum sets of characters needed for identification (reviewed by Morse, 1975; Preece, 1976), but there is little information on what size these sets must actually be in practice.

We have over some years accumulated a good deal of empirical information on this question, and most of it has been taken from diagnostic keys for vascular plants. Some data have also been collected from keys to other organisms and from some other diagnostic devices, enough to suggest that very similar findings obtain for other living creatures. We present this information here, and discuss its significance, and also take into account one example from a key to artificial symbols, i.e. man-made glyphs.

The ratio between the number of characters in a key, m, and the number of taxa, q, is very seldom as low as the theoretical minimum. The number of taxa that can be keyed out obviously cannot be greater than the product of the numbers of states for the characters. For example, for qualitative characters (where each character has two states), $m = 5$ characters allows the separation of $q = 2^5 = 32$ different taxa. Yet this is seldom achieved in practice: usually m has to be at least as great as q. It should be noted that we are here considering the number of characters required to make the entire key, not the number that have to be determined on the particular specimen that is being keyed out. Also, because of the relation to information theory, it is convenient to convert multistate characters to their quantitative (binary) equivalents wherever possible.

There are two main reasons why m has to be greater than the theoretical minimum:

1. Characters are usually not employed repeatedly in the same key.
2. It is usually advisable to employ more than one character in constructing a workable couplet.

The first reason explains why m is commonly almost as large as q; in a dichotomous key (or in a key whose polychotomies have been converted into dichotomies), there are $q - 1$ couplets in a simple key (where each taxon is keyed out only once); therefore, if each couplet requires a different character, $m = q - 1$. In complex keys, where the same taxon or taxa may key out at more than one place, this relationship does not hold, but for most keys there are few such taxa, so the discrepancy is usually minor.

We may therefore view the excess of m over the theoretical minimum, up to the value of $m = q - 1$, as being broadly due to shortage of characters that can be used repeatedly, and over that figure as being due to the need to employ several characters in a couplet as an insurance against character-inconstancy, difficulty of observation, or missing information in the specimen (e.g. plant not in flower).

Characters are not always discriminatory, for often a character is added, perhaps just to one half of a couplet, to help the user to know if he is on the right path. These strictly superfluous, but nevertheless useful, characters were sometimes difficult to accommodate in our analyses and we have often made rather arbitrary decisions about them. A rough estimate from a wide range of *Flora Europaea* keys used to identify herbarium material indicates that to about a third of all the questions asked one cannot come to a clear decision, and must respond with "I don't know", "Perhaps", "Not applicable to this specimen" etc. It is rare to go through a key of more than half a dozen dichotomies without having to retrace one's steps and try other alternatives.

It is evident that the value of m will depend on: (1) the available knowledge of the group and whether the taxonomist finds, and wants to make use of, characters that can be used repeatedly; (2) the variability of characters in the taxa; and (3) the standard of accuracy that the key is to achieve. The greater each of these is, the greater will m need to be. It is perhaps surprising that the relations for vascular plant keys are so regular (*see below*) considering the great variation one would expect in (1), (2) and (3). It is not so surprising that for the other organisms the relations are much less regular.

MATERIALS AND METHODS

The keys and other devices studied were mostly taken from three sources, Volumes 1 and 2 of *Flora Europaea* (Tutin *et al.*, 1964, 1968), *Flora of the British Isles* (Clapham *et al.*, 1962) and *Bergey's Manual of Determinative Bacteriology* (Buchanan and Gibbons, 1974). Some other sources are mentioned individually later. It is hardly necessary to observe at the outset that many taxa contain only one subordinate taxon so no key is necessary, and therefore there are no data for $q = 1$. We aimed in each case to estimate the number of characters m required to construct a key for q taxa, but this soon led to problems of defining a character. In this study a character has been a broad property, such as "leaf-shape" or "flower-colour", which may contain numerous different character-states. "Petal-length in mm" was taken as one character but "petal-length in relation to calyx-length" was taken as a second character.

Different characters contain very diverse amounts of information, as is shown by the varied numbers of character-states. It would have been best if information could have been estimated in the usual way in information theory as the number of binary alternatives ("bits"). However, it is seldom possible to convert keys into a form where only binary alternatives, or two-state characters, are used at each couplet. It is easier to count the number of character-states, g, of each character i. In order to extend the study on these lines, a more careful study was made of a limited selection of keys from Clapham *et al.* (1962), and 10 suprageneric keys and 40 infrageneric keys were chosen at random. The former were keys to genera within families, the latter were keys to species within genera (occasionally including a few hybrids, subspecies or varieties). The selections appeared to be reasonably representative of large and small keys and of "easy" and "difficult" groups.

The properties tabulated were as follows:

1. The number of taxa in the key, q.
2. The number of couplets in the key, c. In cases where a division was given into three alternatives, this was counted as two dichotomous couplets because they could always be rewritten into this form.
3. The number of characters used in the key, m, taken in the broad sense.
4. The number of diagnostic character-states or decisions in the key, d, equivalent to the number of binary decisions that have to be made in order to use the key. A character i with g_i states requires one to make $g_i - 1$ decisions upon that character. For example, with the character "flower-colour" and the four states "red", "yellow", "blue" and "white", one must be able to make the decisions "red"? "yellow"? "blue"? The answer "no" to each of these implies the specimen is white if the alternatives are mutually exhaustive. In terms of information theory, there are three binary pieces of information ("bits") in the character, and the character could be re-scored as three binary (positive-negative or presence-absence) characters, but not as less than three binary characters. In the m characters of the key, therefore, the number of such binary decisions is the sum $\sum^{m} (g_i - 1) = d$. This, therefore, is one measure of the total information in the key.
5. The minimum number of diagnostic decisions that were required to use the key, and this is referred to as d_{min}. In estimating this the characters were chosen so that the fewest decisions were needed even though they might be inconvenient in practical use, e.g. "annual/biennial", "perennial" was chosen if it gave a smaller value for d_{min} than another character that could be much more readily determined from a specimen. This value of d_{min}, therefore, is roughly equivalent to the smallest amount of information on which a key can be built excluding such practical considerations. It

was always larger, however, than the theoretical minimum required to separate q taxa, because the theoretical minimum pre-supposes that one can find characters that can be repeatedly used in different parts of the key, and such characters are in fact quite difficult to discover.

The determination of these various quantities led to a number of difficulties. The number of taxa, q, and the number of couplets, c, were straightforward to count, but in the keys studied here c was always close to $q - 1$, so little additional information was obtained from c.

The number of characters in a broad sense, m, was also quickly determined in all but the longest keys. The estimation of d was a good deal more difficult. This was partly due to the need to keep count of the number of alternative states for each character as it was noted, and partly due to uncertainty as to just how to count the number of states in complex or quantitative characters. For example, how many states are represented by the statements "red", "reddish-purple", "blue or purple", "violet" and "dusky-red"? Again how many states are given by leaf-lengths stated as "10–17 mm, sometimes up to 20 mm", "7–12 mm" and "50–70 mm"?

In these cases the number of separate questions that the user must pose was taken as the guide unless near synonyms were evidently present. In the first, therefore, the user must be able to say if a colour is red, reddish-purple, blue, purple, or dusky, but "violet" could reasonably be considered a synonym of "purple". In the latter, the user must in principle be able to decide if leaves are greater or less than each of the seven lengths 7, 10, 12, 17, 20, 50 and 70 mm, even though no plants have lengths between 20 and 50 mm. It was assumed that the inclusion of all these did imply they were of significant value to the user. Rather arbitrary choices had to be made in doubtful cases.

The work entailed in estimating d was considerable, and set a limit to the sample that could be studied in a reasonable time: in the worst cases, the characters had to be tabulated with all their states, which was quite laborious.

The minimum number of decisions, d_{min}, was also difficult to estimate. Possible alternative characters were numerous, and it would have been impracticable to tabulate all the combinations to ensure that the absolute minimum had been found. However, it was soon noted that d_{min} was usually close to c (and therefore also usually close to q), that is, each couplet required a different binary decision. Departures from this had two causes. First, if a given decision can be used more than once in a key, d_{min} is less than c. Second, if a couplet cannot be worked without several decisions, then d_{min} is greater than c; this happens with couplets of the type "stems woody, with alternate leaves, or flowers blue or purple" *vs.* "not as above". In a few cases it would have been possible by consulting the descriptions of the plants to reduce d_{min} in such cases, but this was not done, because the study was intended to be upon actual keys rather than ideal ones.

These investigations were extended to bacteria, which are often difficult to identify even when the taxonomy is well worked out. Fifty keys from *Bergey's Manual of Determinative Bacteriology* (Buchanan and Gibbons, 1974) were examined, including keys to both high and low taxonomic ranks. The number of taxa was taken as the number of groups actually separable with the key (the termini sometimes listed several species that could not be distinguished by the key). A number of diagnostic tables were also examined, but relationships between m, d and q were much less regular, so that only a few general observations are made on these.

Some zoological keys were also examined from a variety of sources, mostly for arthropods. In some of these keys numerous states are used within a character such as chaetotaxy, e.g. segments 2–8 with "chaetae numbering 2,2,4,4,4,4,0 or 2,4,4,4,4,4,0" *vs.* "different from the above". Clearly, unless one is told one may ignore some segments, there is a minimum of eight observations to make: this can inflate d considerably.

The findings are presented as scatter-diagrams (Figs 1–10), to which linear regressions on q have been fitted where appropriate, and as a table summarizing the main ratios (Table I).

RESULTS

Vascular plants

The scatter diagrams for the 104 keys from *Flora Europaea* are shown in Fig. 1. It is surprising how many characters are used in these keys. Taking characters in a broad sense, we find m/q is about 1.

The relationship is somewhat curved, so that at high values of q the ratio is somewhat less. For example, a typical small genus, *Hippocrepis*, has $q = 10$ and $m = 15$; a typical medium-sized genus, *Ononis*, has $q = 49$, $m = 49$; and a typical large genus, *Silene*, has $q = 166$ and $m = 109$. There is no very obvious explanation for this curvature; transformation of q to log q did not make the trend linear.

There are a few points that do not fit the trend well. *Alchemilla* and *Astragalus* have very low m/q ratios (0·40 and 0·53, respectively). *Alchemilla* is apomictic, and is treated taxonomically in a different way from most other genera. It cannot be said that the small number of characters is necessarily due to the recognition of somewhat arbitrary taxa, although it will be realized that if artificial groupings

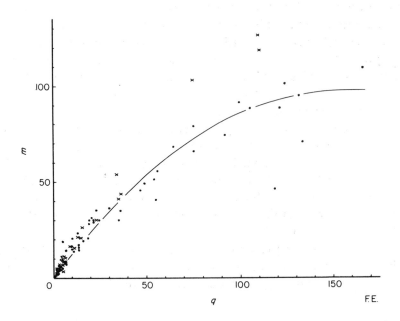

FIG. 1 Scatter-diagram of the number of characters, m, against the number of taxa, q, for 104 keys from *Flora Europaea* (Tutin *et al.*, 1964). 146 taxa were studied: 42 were monospecific genera for which no keys are required. The keys to genera within families are symbolized with a cross (×), those to species within genera with a full point (·). Most of the keys to species were from the Leguminosae and Rosaceae. Since the relationship of m to q appears curved, a quadratic regression line has been fitted with the constants $m - 1·577 + 1·286q - 0·0043q^2$. (This regression is inappropriate for high q because this curve then begins to descend again.)

are used for convenience in a genus with a very confused pattern of variation, this will permit keys with few characters (e.g. as was observed earlier, if each combination of five characters is treated as a taxon, one may be able to establish $q = 2^5 = 32$ "taxa").

Astragalus on the other hand is not apomictic, but it is unusually difficult taxonomically, being a very large genus with many very incompletely known sections.

The results from the keys in *Flora of the British Isles* (Clapham *et al.*, 1962) are shown in Figs 2–4. There is no evidence of curvilinearity, perhaps because there were few large keys. In contrast to *Flora Europaea* the ratio m/q is about 2 instead of about 1 over the same ranges of q. The characters were scored in about the same way for both, so the difference in slope is not due to differences in this. Possibly the need to keep the *Flora Europaea* keys concise has led to paring down the number of characters to a minimum.

Higher rank keys may have a higher m/q ratio, as the points for keys to genera lie higher on the scatter-diagram in Fig. 1, but this does not seem so for Fig. 2. Further, a count of the key to families of monocotyledons (Hutchinson, 1934) showed $q = 68, c = 125, m = 152$ and $d = 242$ (d_{min} was difficult to estimate, but was close to c). It is not clear, therefore, whether rank affects the m/q ratios in general.

The ratio of d/q for the keys in *Flora of the British Isles* is about 2·8, reflecting an average of roughly $2\frac{1}{2}$ states per conventional character (Fig. 3). The values of d_{min} are always close to c and to $q - 1$ (Fig. 4), so it is seen that about one binary character per taxon is the minimum practical number.

Bacteria

The points in the diagrams from data on bacterial keys are notably more scattered for m and d (Figs 5 and 6). In part this may reflect the less uniform taxonomic treatment of bacteria. In part, too, it reflects less natural or even frankly arbitrary groupings. Thus in the key to species of *Clostridium*, four of the five major sections are simply convenient species-groups based on terminal *vs.* subterminal spores and on hydrolysis of gelatin, and such keys show values of m and d that are less than q and approach the minimum theoretical lower limit given by $q = 2^m$. The m/q ratio still averages about 1·2. Since most characters are presence-absence characters, d is not greatly above m.

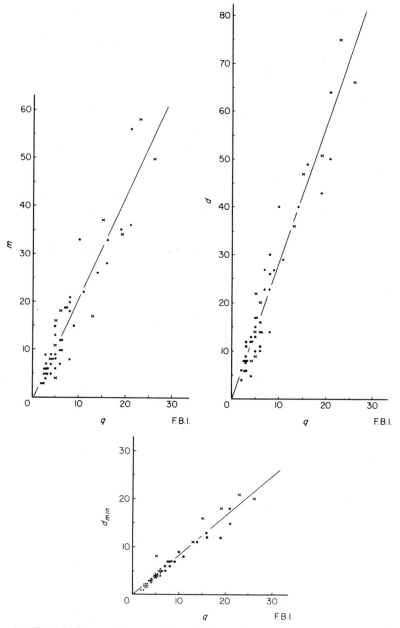

FIG. 2. (*Top left*) Scatter-diagram of *m* and *q* for 50 keys from *Flora of the British Isles* (Clapham *et al.*, 1962), with the regression line of *m* on *q*. Symbols as in Fig. 1.

FIG. 3. (*Top right*) Scatter-diagram of *d* on *q* for the same keys from Clapham *et al.* (1962) that are plotted in Fig. 2. Conventions as in Figs 1 and 2.

FIG. 4. (*Bottom*) Scatter-diagram of d_{min} and *q* for the same keys from Clapham *et al.* (1962) that are plotted in Fig. 2. Conventions as in Figs 1 and 2.

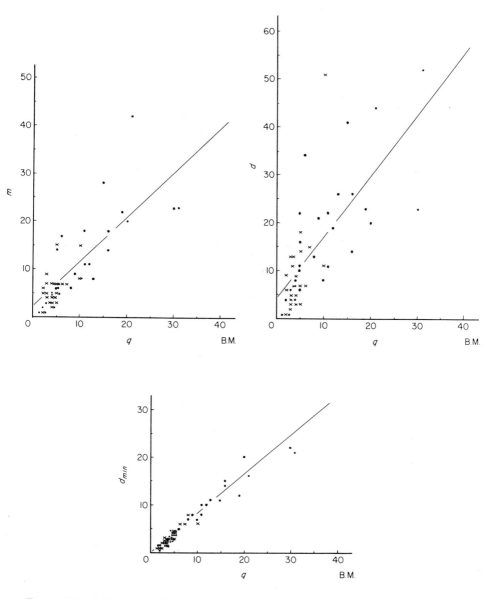

Fig. 5. (*Top left*) Scatter-diagram of *m* and *q* for 50 keys to bacteria (Buchanan and Gibbons, 1974). Conventions as in Fig. 2 except that the symbol × includes keys to ranks above family.

Fig. 6. (*Top right*) Scatter-diagram of *d* and *q* for the keys used in Fig. 5, with conventions as in Fig. 5.

Fig. 7. (*Bottom*) Scatter-diagram of d_{min} and *q* for the keys used in Fig. 5, with conventions as in Fig. 5.

There is no evidence of curvilinear relationships, or of a different behaviour for keys to higher taxa. Nevertheless, the value of d_{min} is close to c and to $q - 1$ as a rule (Fig. 7).

Some counts of m, q and d on 50 diagnostic tables from Buchanan and Gibbons (1974) showed rather similar behaviour to those found with keys. The m/q and d/q ratios averaged about 1·8 and 2·6 respectively, but the points in scatter-diagrams were even more widely scattered. It was not feasible to determine d_{min} for these tables.

Animals

A series of 26 zoological keys were taken from Kaston and Kaston (1953), Longfield (1937), Levi (1974), Platniels (1974), Mayer (1974), Scullen (1972), Dawes (1968) and Jacobson and Kirstner (1975). Most were of arthropods, but fishes and trematodes were included. Very similar relationships between m, q and d were found (Figs. 8–10, Table I), although there was much scatter. The high

TABLE I.   *Means and regressions of variables of keys.*

	q	c	m	d	d_{min}
Keys in *Flora Europaea*					
Means (n = 104)	25·07	29·46	25·20		
Keys in *Flora of the British Isles*					
Means (n = 50)	8·12	7·12	16·84	22·90	6·64
Regression on q; a			−0·40	−0·11	−0·05
Regression on q; b			2·13	2·83	0·82
Correlation with q			0·95	0·97	0·97
Keys in *Bergey's Manual*					
Means (n = 50)	7·80	6·88	9·38	14·20	6·26
Regression on q; a			2·03	4·16	−0·10
Regression on q; b			0·94	1·29	0·82
Correlation with q			0·80	0·71	0·97
Zoological keys					
Means (n = 26)	12·88	11·96	21·58	40·73	11·31
Regression on q; a			−0·80	−10·42	−0·27
Regression on q; b			1·74	3·97	0·90
Correlation with q			0·82	0·84	0·97

values of d were partly due to the inclusion of much detailed information on colour patterns and chaetotaxy. Again, d_{min} is close to c and to $q - 1$ (Fig. 10).

Man-made symbols

A striking contrast to keys for living organisms is shown by a celebrated key (Shepherd, 1971) to man-made symbols or glyphs. Here the values of q, m and d are respectively 434, 98 and 179 (with a d_{min} of 179, as no couplets use more than one character). Although we have found no other keys to glyphs, evidently the ratios can be far lower than those usual with living organisms. The reasons evidently lie in the nature of glyphs: they are artificial, they can be completely defined by a few characters (and indeed were created to be both as simple and diagnostic as possible), and classes of glyphs are artificial groupings. In consequence, almost all combinations of character-states can and often do occur.

Despite these considerations, there is presumably some restriction on the possible combinations of character-states, because glyphs are influenced by other glyphs, if only because a glyph that is too like an established one will not be adopted. But the restrictions appear to be slight. It may be fanciful to think of glyphs evolving to fill social niches, but if they do evolve they have not yet evolved enough to lead to the pronounced character-correlations found in living organisms that prevent the d/q ratio approaching the theoretical lower limit given by $q = 2^d$.

DISCUSSION

There are not many published data on m/q ratios. In bacteriology it was noted that the ratio was often about $1\cdot2$ (Sneath, 1969), and that, even with computer identification methods, the proportion of successful identifications began to fall markedly if the ratio was reduced below $0\cdot6$ by omitting tests. In yeasts, Barnett and Pankhurst (1974) were able to construct keys on presence-absence tests with mean m/q of $0\cdot59$, but a few taxon pairs were then not separable. In practice, confirmatory tests had be be added to raise the ratio to about $0\cdot85$.

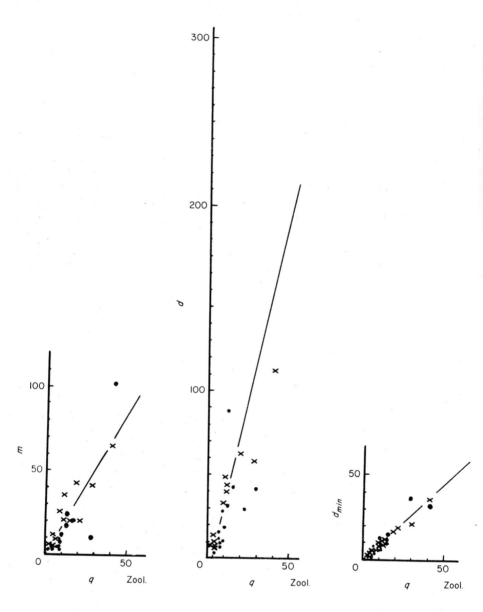

FIG. 8. (*Left*) Scatter-diagram of *m* and *q* for 26 zoological keys from various sources (*see text*). Conventions as in Fig. 2.

FIG. 9. (*Centre*) Scatter-diagram of *d* and *q* for the zoological keys used for Fig. 8. Conventions as in Fig. 2.

FIG. 10. (*Right*) Scatter-diagram of d_{min} and *q* for the zoological keys used for Fig. 8. Conventions as in Fig. 2.

There is some evidence that fewer characters are needed for diagnostic tables when the unknowns are compared with the taxa by computer methods than when this is done manually (Sneath, 1969). It is not known whether computer-made keys of the kind described by Pankhurst (1970) require fewer characters than those made by hand, though the saving is probably not very large. We have here the difficulty of testing the efficiency of keys in an expeditious way: it is possible that it would be worth developing a computer method of identifying by a key large numbers of specimens (or simulated specimens) to investigate this problem. Nor is it clear whether genera with very numerous species require more characters in keys to genera within a family than do genera with a few species (the data on *Flora Europaea* give no evidence of any marked differences).

It might be of interest to study whether there is any connection between the regularities in key information content and the "Hollow Curves" of Willis (1922), which seem to be found on tabulating the number of taxa containing 1, 2, 3 . . . subordinate taxa.

The main conclusions from this study are that almost all workable keys (and probably also diagnostic tables) have a ratio of the number of characters to taxa that is well over 1, especially if characters are expressed as binary ones, or if character-states are counted. Of this ratio, the portion up to $1·0$ usually represents the minimum number of characters or character-states that gives separation between all taxa, whereas that above $1·0$ reflects mainly the additional information used to gave added reliability to the key. It is uncommon for d_{min} to be very different from $q - 1$ or c, so that as far as the number of characters is concerned, it seldom matters whether taxa are keyed out one at a time or are successively divided at each couplet into subsets of approximately equal size.

Just as the experienced user of keys comes to feel suspicious of any large key that keys out each taxon only once, so he should perhaps be wary of compiling, or placing too much faith in, keys where the ratio of the number of characters to taxa is 1 or less. It appears that with many patterns of natural variation one cannot expect to achieve workable identification with anything like the theoretical minimum of characters—a point that may have implications in other fields such as ecology, medicine etc. The observation of Gorry and Barnett (1968) that 31 attributes were needed for diagnosis of 35 diseases of the heart is in keeping with this.

There are exceptions to the above generalization for very difficult or poorly known taxa, where high or low ratios may to some extent indicate exceptional difficulty in finding good characters or reflect groupings that are frankly artificial. No very pronounced difference is found with regard to different levels of rank of the categories, but when the number of taxa is very large the ratio tends to be lower. This may imply difficulty in finding enough characters, or perhaps it is that keys to small numbers of taxa are apt to be over-elaborated out of a desire to construct a really foolproof key. Nevertheless, one would expect a falling off with large numbers of taxa, because it seems improbable that the quarter-million species of flowering plants require a quarter-million different characters to key them out. It may be significant, therefore, that the identification system for angiosperm families of Hansen and Rahn (1969), though it is a punched-card polyclave, allows determination of 411 families by 172 cards, equivalent to a d of about 140.

REFERENCES

Barnett, J. A. and Pankhurst, R. J. (1974). "A New Key to the Yeasts: A Key for Identifying Yeasts Based on Physiological Tests Only." North-Holland, Amsterdam.

Buchanan, R. E. and Gibbons, N. E., eds. (1974). "Bergey's Manual of Determinative Bacteriology," 8th edition. Williams and Williams, Baltimore.

Clapham, A. R., Tutin, T. G. and Warburg, E. F., (1962). "Flora of the British Isles," 2nd edition. Cambridge University Press, Cambridge.

Dawes, B. (1968). "The Trematoda with Special Reference to British and Other European Forms." Cambridge University Press, Cambridge.

Gorry, G. A. and Barnett, G. O. (1968). Sequential diagnosis by computer. *J. Amer. Med. Ass.* **205**, 849–854.

Hansen, B. and Rahn, K. (1969). Determination of angiosperm families by means of a punched-card system. *Dansk bot. Ark.* **26**, 1–44.

Hutchinson, J. (1934). "The Families of Flowering Plants. II. Monocotyledons." Macmillan, London.

Jacobson, H. R., and Kistner, D. H. (1975). The natural history of the myrmecophilous Tribe Pygostenini (Coleoptera: Staphylinidae). Section 12. A manual for the identification of the Pygostenini. *Sociobiology* **1**, 201–335.

Kaston, B. J. and Kaston, E. (1953). "How to Know the Spiders." W. C. Brown, Dubuque, Iowa.

Levi, H. W. (1974). The orb-weaver genus *Zygiella* (Araneae, Araneidae). *Bull. Mus. comp. Zool.* **146**, 267–290.

Longfield, C., (1937). "The Dragonflies of the British Isles". Warne, London.

Mayer, G. F. (1974). A revision of the cardinalfish genus *Epigonus* (Perciformes, Apogonidae) with descriptions of two new species. *Bull. Mus. comp. Zool.* **146**, 147–203.

Morse, L. E. (1975). Recent advances in the theory and practice of biological specimen identification. *In* "Biological Identification with Computers" (R. J. Pankhurst, ed.), pp. 11–52. Academic Press, London and New York.

Pankhurst, R. J. (1970). A computer program for generating diagnostic keys. *Computer J.* **12**, 145–151.

Pankhurst, R. J., (1975). Identification methods and the quality of taxonomic descriptions. *In* "Biological Identification with Computers" (R. J. Pankhurst, ed.), pp. 237–247. Academic Press, London and New York.

Platnick, N. (1974). The Spider Family Anyphaenidae in America North of Mexico. *Bull. Mus. comp. Zool.* **146**, 205–206.

Preece, D. A. (1976). Identification keys and diagnostic tables. *Math. Scientist* **1**, 43–65.

Scullen, H. A. (1972). Review of the genus *Cerceris* Latreille in Mexico and Central America (Hymenoptera: Sphecidae). *Smithsonian Cont. Zool.* No. **110**.

Shepherd, W. (1971). "Shepherd's Glossary of Graphic Signs and Symbols." J. M. Dent, London.

Sneath, P. H. A. (1969). Computers in Bacteriology. *J. Clin. Pathol.* **22**, suppl. 3, 87–92.

Sneath, P. H. A. and Sokal, R. R. (1973). "Numerical Taxonomy: The Principles and Practice of Numerical Classification." W. H. Freeman, San Francisco.

Tutin, T. G., Heywood, V. H., Burges, N. A., Valentine, D. H., Walters, S. M. and Webb, D. A., eds. (1964). "Flora Europaea," Vol. 1 (1964), Vol. 2 (1968). Cambridge University Press, Cambridge.

Willis, J. C., (1922). "Age and Area. A Study of the Geographical Distribution and Origin of Species." Cambridge University Press, Cambridge.

Chapter 6

The Classification of Crop Plants

P. F. PARKER

INTRODUCTION

In utilizing and cultivating plants, primitive man had to learn to recognize and differentiate between the various types. The necessity of identifying plants, first in the wild state, and later under cultivation, slowly led to a rudimentary knowledge of grouping and classifying plants.

The accumulation of this knowledge over the centuries finally led to taxonomy, the earliest branch of botany. At the beginning then, taxonomy was founded on man's knowledge of useful wild plants, and of cultivated plants. Only much later, as taxonomy developed, did interest shift to the classification of plants other than those directly useful to man. As interest in wild species developed, so that in cultivated plants has diminished steadily, until today it is a relatively neglected field.

Against this background of the taxonomist's disinterest, we have to place the fact that since the early part of the twentieth century, crop botanists and plant breeders have used the newer disciplines of genetics, cytology and, more recently, biochemistry, for studies of cultivated plant species and genera. These studies were designed to give a clearer insight into the origins and nature of individual crop plants. In a few groups, such studies by non-taxonomists have resulted in monographs of such genera as *Nicotiana* (Goodspeed, 1954), *Solanum* (Dodds, 1962; Hawkes, 1958) and *Cicer* (Van der Maesen, 1972), among others. More often, however, the result of such work has been a series of papers by different groups of workers;

papers which often remain uncollected and largely unknown, except by specialists.

Modern taxonomists, such as Ornduff (1966), Nordberg (1967) and Persson (1974), have taken the techniques first used in the study of cultivated plants and applied them to biosystematic studies of natural populations with considerable success, but a general lack of interest in crop plants has, with few exceptions, continued. It may be useful to consider the reasons advanced for this by some crop plant evolutionists (Heiser, 1968; Harlan and de Wet, 1971).

Although there is no absolute distinction between the systematics of cultivated as opposed to wild plants, there are differences between them which may lead to problems. One of the most important of these is that plants under cultivation do not as a rule have a natural population structure. They grow in simple communities, and conditions of growth are more or less modified in their favour by man. There is often great diversity present also, and variation patterns may be unlike those found in the wild. Sources of this variation can be attributed to hybridization and introgression between cultivated and wild species of the same chromosome number as in *Lactuca* (Lundquist, 1960); vegetative propagation of hybrids and backcrosses between cultivated and wild species of different chromosome number, as in tuberous *Solanum* (Ugent, 1968); or the early formation of new complex polyploid cultivars from sequential interspecific hybridization as in *Triticum* (Riley, 1965).

Cultivar groups are also often continuously distributed over extensive land areas, or discontinuously distributed throughout the tropics. In these situations adequate sampling is difficult, leading to a slow accumulation of information on the distribution of species and races, and consequently a considerable delay in the classification of the full range of variation. Anderson (1954) pointed out that, in nearly all cases, herbarium material is inadequate and unrepresentative, and, with regard to the fruits in particular, it is difficult to assemble a collection of good specimens. This is, however, no real obstacle, as the work of Anderson and his associates have demonstrated with *Zea mais* L. (Anderson, 1941, 1947; Anderson and Cutler, 1942), *Persea americana* L. (Anderson 1954) and the grain species of *Amaranthus* (Sauer, 1950). No obstacles either are the difficulties of nomenclature which exist where a crop and its relatives have been poorly or superficially classified by non-taxonomists, or where previous specialists have disagreed about specific or generic boundaries. Such problems can lead to a considerable amount of extra work; however, they are relatively easily overcome compared to the problem raised by Stuart (1974) concerning the time scale at the cultivar level.

In developed and developing countries, both continuous replacement and loss of cultivars are important factors, and not only with those crops that are annuals or treated as annuals. In developing countries, land races are disappearing as they are replaced by newly developed local or imported cultivars. Except in the case of some major crop plants, these land races have not been fully classified and their loss is absolute. A different state of affairs exists in the developed countries. Here there is a constant production of new cultivars, and withdrawal of old ones. The problem is how to classify what may be a constantly changing spectrum of variation. It is possible to do this given the resources, but where a satisfactory system has not been devised or agreed upon, or where there is the introduction of completely new morphological characters from species not previously involved in the development of a cultivar group (Knight and Keep, 1964), a taxonomist engaged in this work is often fighting a losing battle, and must inevitably fall far behind.

I feel, however, that none of these difficulties are at the root of the problem. Despite recognition of the fact that a sound knowledge of the wild relatives of cultivated taxa is essential to the interpretation and full understanding of the origins of cultivated plants, most taxonomists are at present more concerned with the broader taxonomic situation.

The floras of large areas of the world have not yet been adequately described and classified, and other areas have not been treated for over 100 years. In Europe, taxonomists are working on the Floras of Turkey (Davis, 1965 *et seq.*) Iran (Rechinger, 1965 *et seq.*), the Zambesi Basin (Exell and Wild, 1960 *et seq.*) and tropical East Africa (Turrill and Milne-Redhead, 1952 *et seq.*). This work is being done on slim budgets, and with a great deal of voluntary taxonomic assistance (Davis, 1965; Abilio Fernandes, personal communication).

Turkey, Iran and other near-eastern and middle-eastern countries were a cradle of Old World civilization. Here many of our cultivated plants originated, and here their progenitors still exist, yet with presently available resources a period of some 15–20 years is needed to produce basic modern Floras for only part of this area. Floras such as these are urgently needed to replace *Flora Orientalis* (Boissier, 1867–1888) and provide a basis for further detailed distributional and genetical work on economically important genera that occur in this area.

In the tropics the situation is worse. *The Flora of Peru* (MacBride, 1936 *et seq.*) is still incomplete; *Flora Brasiliensis* (von Martius, 1840 *et seq.*) was never completed; *Flora Neotropica* (Howard *et al.*, 1968 *et seq.*) has only just commenced. The tropics are rich in plants that may yield drugs or provide new food crops on a local or global basis. Taxonomists are still far behind with the production of the descriptive and classificatory work, and also with the distributional studies that are necessary both to provide a basis for phytochemical and agronomic surveys and to gain a sufficient knowledge of disappearing wild species so that theories appertaining to their phylogenetic relationships may be developed.

This slow progress in the face of the rapid destruction of tropical and subtropical vegetation in particular, and this overriding objective of collating evidence relating to the long-term evolutionary relationships of the world flora, are basic reasons for the disinterest in cultivated plants shown by the majority of professional taxonomists. Given this global perspective, the classification and understanding of the origins of a few hundred crop plants pale into insignificance.

CLASSIFICATION ABOVE THE SPECIES LEVEL

Despite the difficulty of interesting taxonomists in this field, the classification of the more important crop plants has gone ahead. Distributional, morphological, ecological, genetic and cytotaxonomic studies of cultivated taxa and their wild relatives, supported where possible by archaeological data, have produced a mass of information which has to be synthesized and refined into a classification reflecting all that is known about a group. It is this final stage which has not always been satisfactory from the point of view of practitioners in the field of crop plant evolution, and they have tended to reject the classical system as inappropriate (Harlan and de Wet, 1971). However, any system, whether a "morphological" one, with its groupings based on a sufficiently close resemblance among the phenotypes of the organisms concerned, and a sufficiently great distinction between phenotypes of one group and the phenotypes of other groups, or a "biological" (or "genetic") one, based on genome relationships and sterility barriers, and thus gene pool limits, should, given *the same totality of information*, lead to broadly similar results, although of course they have an entirely different purpose.

A classification system which emphasizes genetical relationships and discontinuities has considerable advantages for all who work with cultivated plants and their relatives, especially from the point of view of sources of genetic material. However, the gene-pool concept as proposed by Harlan and de Wet (1971) (Fig. 1.), although excellent when used to show general relationships, is not a classification as such. It is, moreover, a system which can only be applied when a great

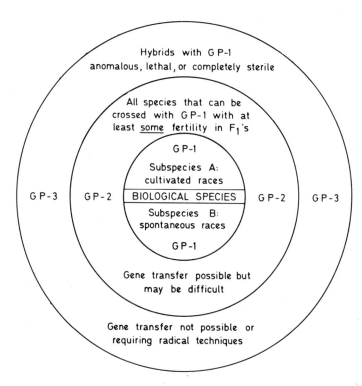

FIG. 1. The gene-pool concept proposed by Harlan and de Wet (1971). Satisfactory only after all the taxa concerned have been intensively studied and satisfactorily classified.

deal of information at all levels has been gained about the group under examination. This limits its application to a considerable extent, and is why most taxonomists, although taking into account all the experimental information available to them, still use a "morphological" system ultimately based on distributional and "morphological" criteria. (Characters used to determine resemblance may include those which are cytological, anatomical and embryological, yet they may still be morphological in the broadest sense).

The major groups of cultivated plants are still lacking fundamental information even on such matters as distribution patterns of individual taxa. However, the volume of information available on all aspects of the biology of crop plants, and to a lesser extent their wild relatives, makes them ideal subjects for the application of systems expressing their basic similarities and differences, using standard botanical nomenclature at least as far as the species level. A consideration of some recent examples may be apposite.

The earliest classifications of the genus *Triticum* have been fully reviewed by Percival (1921). His classification concerned only *Triticum* as it was then understood, and proposed two species to cover the total range of variation, the diploid *T. aegilopoides* Bal. with one race, and *T. dicoccoides* Körn. with 11 races, some tetraploid, some hexaploid (Table I). Soon afterwards, Eig (1929), and later Kihara (1954), published classifications of the genus *Aegilops*, a genus accepted as separate from *Triticum* by Hackel (1887).

TABLE I.   *The classification proposed by Percival (1921) for the cultivated wheats*

Species		Race	
I.	*T. aegilopoides* Bal.	I.	*T. monococcum* L.
II.	*T. dicoccoides* Körn.	II.	*T. dicoccum* Schuble
		III.	*T. orientale* Perc.
		IV.	*T. durum* Desf.
		V.	*T. polonicum* L.
		VI.	*T. turgidum* L.
		VII.	*T. pyramidale* Perc.
		VIII.	*T. vulgare* Host
		IX.	*T. compactum* Host
		X.	*T. sphaerococcum* Perc.
		XI.	*T. spelta* L.

The importance of wheat as a crop plant has led to extensive genome analysis in the *Aegilops–Triticum* group (for reviews see Morris and Sears, 1967; Sears, 1969). These studies showed that hybridization between the two genera was easy, resulting in the identification of three basic genomes (A, B and D) in wheat, but they failed to establish a reasonable homology between given genomes in polyploids and those of known diploid species. These results led Kihara (1954) to propose that many of the sets found in polyploid species of *Aegilops* were modified genomes, only partly homologous with those at the diploid level. Zohary and Feldman (1961), reviewing the genus *Aegilops*, pointed out that, as in other polyploid pillar complexes, specific boundaries were sharp, and ecological ranges narrow at the diploid level, whereas at the polyploid level they were indistinct. They also pointed out that Kihara's modified genomes existed both at the diploid and the polyploid level in three cytological species clusters, each cluster having one unaltered common genome, plus one (or two) modified differential genomes. Another characteristic of the polyploids was that some exhibited combinations of traits found among three or more diploid species groups.

As natural interspecific hybridization is common at the tetraploid level, Zohary and Feldman (1961) have suggested that the polyploids are to be considered derivatives of only a restricted number of initial amphidiploid combinations. When hybridization was between plants with a common genome, this common genome served as a buffer; substitutions and chromosomal alterations occurred mainly between the unshared chromosome sets, and resulted in the establishment of a modified genome alongside an unaltered one. This, and work by Riley and Chapman (1958) on the genetic control of homoeologous pairing (pairing between similar chromosomes from different chromosome sets) in *Triticum* polyploids, has considerably assisted in the understanding of relationships within the group.

Recent doubts as to whether *T. speltoides* (Taush.) Gren. donated the "B" genome or not, remain unresolved. Evidence presented by Shands and Kimber (1974) and Rubenstein and Sallee (1974), suggest that the "modified B" genome in *T. timopheevi* (Zhuk.) Zhuk. var. *timopheevi* is very similar to the "B" (now designated "S") genome of *T. speltoides*, and different from the "B" genome in other wheats. Larsen (1974) suggests that the "B" genome is probably a hybrid genome from various diploid species, and not to be found any longer in a single donor species. The occurrence of introgression from the diploids *T. speltoides* (S genome), *T. longissimum* (Schweinf. & Muschli) Bowden (S¹ genome), *T. dichasians* (Zhuk.) Bowden (C genome) and *T. umbellulatum* (Zhuk.) Bowden (Cᵘ genome), into the tetraploid *T. turgidum* L. var. *dicoccoides* (Körn.) Bowden (ABᵈ genomes) in different parts of its range (Vardi, 1974), lends support to this hypothesis.

With the discovery that at least one, and probably two of the genomes of the polyploid wheats had come from the genus *Aegilops*, it became obvious that a taxonomic revision would have to be undertaken, and a classification produced that would indicate relationships at all levels.

Jakubziner (1958) reclassified the *Triticum* species and recognized three diploid, twelve allotetraploid, six allohexaploid and one auto-allohexaploid species (Table II). He was criticized for giving new names to species that already existed, and for splitting species unnecessarily.

TABLE II. *The classification of Jakubziner (1958) for the wheats only, and that of Bowden (1959) which incorporated the whole of the genus* Aegilops *into the genus* Triticum.

JAKUBZINER (1958)	BOWDEN (1959)
DIPLOIDEA ($2n = 14$)	I. DIPLOID SPECIES. ($2n = 14$)
Wild growing	*T. monococcum* L. (genome AA)
T. boeoticum Boiss.	*T. bicorne* Forsk.
T. urartu Tum.	*T. speltoides* (Tausch) Gren. ex Richter
	(? genome BB)
Cultivated—Glumaceous	f. *speltoides*
T. monococcum L.	f. *ligusticum* (Savign.) Bowden, comb. nov.
	T. comosum (Sibth. & Sm.) Richter
	T. uniaristatum (Vis.) Richter
	T. longissimum (Schweinf. & Muschl. in Muschl.) Bowden, comb. nov.
	T. umbellulatum (Zhuk.) Bowden, comb. nov.
	T. tripsacoides (Jaub. & Spach) Bowden, comb. nov.
	f. *tripsacoides*
	f. *loliaceum* (Jaub. & Spach) Bowden
	T. dichasians (Zhuk.) Bowden, comb. nov.
	T. aegilops P. Beauv. ex R. & S. (genome DD)

TABLE II —*continued*

JAKUBZINER (1958)	BOWDEN (1959)
TETRAPLOIDEA ($2n = 28$) Wild growing *T. araraticum* Jakubz. *T. dicoccoides* (Körn) Schweinf.	II. (a) ALLOTETRAPLOID WHEATS. ($2n = 28$, genome AABB or AAGG) *T. turgidum* L., emend.
Cultivated—Glumaceous *T. timopheevi* Zhuk. *T. paleocolchicum* Men. *T. dicoccum* Schubl.	Groups of cultivars 1. *turgidum* L., *polonicum* L., *dicoccon* Schrank., *durum* Desf., *carthlicum* Nevski, *turanicum* Jakubz., *paleocolchicum* Men., *aethiopicum* Jakubz.
Cultivated—Naked *T. durum* Desf. *T. turgidum* L. *T. turanicum* Jakubz. *T. polonicum* L. *T. carthlicum* Nevski *T. aethiopicum* Jakubz.	2. var. *dicoccoides* (Körn.) Bowden, comb. nov. 3. var. *timopheevi* (Zhuk.) Bowden, comb. nov. f. *timopheevi* f. *zhukovskyi* (Men. & Er.) Bowden, comb. nov. 4. var. *tumanianii* (Jakubz.) Bowden, comb. nov.
HEXAPLOIDEA ($2n = 42$) Cultivated—Glumaceous *T. zhukovskyi* Men. et Er. *T. macha* Dek. et Men. *T. spelta* L.	II. (b) ALLOHEXAPLOID WHEATS. ($2n = 42$, AABBDD) *T.* × *aestivum* L., emend.
Cultivated—Naked *T. aestivum* L. *T. compactum* Host *T. vavilovii* Jakubz. *T. sphaerococcum* Perc.	Groups of cultivars 1. *aestivum* L. 2. *spelta* L. 3. *compactum* Host 4. *sphaerococcum* Perc. 5. *macha* Dek. & Men. 6. *vavilovii* Jakubz.
	II. (c) ALLOPOLYPLOID SPECIES OF INTERSPECIFIC HYBRID ORIGIN. ($2n = 28$ or 42) *T. ovatum* (L.) Raspail *T. triaristatum* (Willd.) Godr. & Gren. *T. kotschyi* (Boiss.) Bowden *T. triunciale* (L.) Raspail *T. cylindricum* Ces. *T. macrochaetum* (Shuttle & Huet. ex Duval-Jouve) Richter *T. crassum* (Boiss.) Aitch. & Hemsl. *T. turcomanicum* (Rosh.) Bowden *T. ventricosum* Ces.
	III. OTHER ARTIFICIAL AND NATURAL INTERSPECIFIC HYBRIDS.

E

TABLE III. *Morris and Sears (1967) modification of the classification of Bowden (1959), and the classification of MacKey (1968) which recreates Aegilops as a genus and defines new genera and sections from the genus Triticum.*

MORRIS and SEARS (1967)		MACKEY (1968)		
Species	Genome formula	Species	$2n$	Genome type and cluster
DIPLOIDS		**Aegilops L.**		
		Polyeides (Zhuk.) Kihara		
T. monococcum L.	A	Ae. umbellulata Zhuk.	14	C^u
T. speltoides (Tausch) Gren. ex Richter	$S(=B?)$	Ae. ovata L.	28	$C^u M^o$
T. bicorne Forsk.	$S^b(=B^b?)$	Ae. triaristata Willd.	28	$C^u M^t$
T. longissimum (Schweinf. & Muschli in Muschli) Bowden	$S^l(=B^l?)$	Ae. recta (Zhuk.) Chenn.	42	$C^u M^t M^{t2}$
T. tripsacoides (Jaub. & Spach) Bowden	$M^l(=B??)$	Ae. columnaris Zhuk.	28	$C^u M^c$
		Ae. biuncialis (Vill.) Vis.	28	$C^u M^b$
T. tauschii (Coss.) Schmal.	D	Ae. variabilis Eig	28	$C^u S^v$
T. comosum (Sibth. & Sm.) Richter	M	Ae. triuncialis L.	28	$C^u C$
T. uniaristatum (Vis.) Richter	M^u	Cylindropyrum (Jaub. et Spach) Kihara		
T. dischasians (Zhuk.) Bowden	C	Ae. caudata L.	14	C
T. umbellulatum (Zhuk.) Bowden	C^u	Ae. cylindrica Host	28	CD
		Vertebrata (Zhuk.) Kihara		
ALLOPOLYPLOID WHEATS		Ae. squarrosa L.	14	D
T. turgidum L.		Ae. crassa Boiss. 4 x	28	DM^{cr}
var. dicoccoides (Körn.) Bowden	AB	Ae. crassa Boiss. 6 x	42	$DD^2 M^{cr}$
Groups of varieties:	AB	Ae. vavilovii (Zhuk.) Chenn.	42	$DM^{cr} S^l$
dicoccon, durum, turgidum,		Ae. ventricosa Tausch.	28	DM^v
polonicum, carthlicum		Ae. juvenalis (Thell.) Eig	42	$DC^u M^i$
		Amblyopyrum (Zhuk.) Kihara		

Species	Genome	Species (synonym)	2n	Genome
T. timopheevi (Zhuk.) Zhuk.	$AB, A\beta,$ or AG	Comopyrum (Jaub. et Spach) Sen. Korch		
var. timopheevi		Ae. comosa Sibth. et Sm.	14	M
var. zhukovskyi (Men. & Er.) Morris & Sears	$AAB, AA\beta,$ or AAG	Ae. uniaristata Vis.	14	M^u
T. aestivum L. em. Thell.	ABD	Sitopsis Jaub. et Spach		
		Ae. speltoides Tausch.	14	$S(=B)$
Groups of varieties:		Ae. bicornis (Forsk.) Jaub. et Spach	14	S^b
spelta, vavilovii, aestivum, compactum, sphaerococcum		Ae. longissima Schweinf. et Muschl	14	S^c
		Crithodium Link		
OTHER ALLOPOLYPLOIDS		Cr. aegilopoides Link	14	A
T. ventricosum Ces.	DM^{vr}			
T. crassum (Boiss.) Aitch.[a] & Hemsl.[b]	$\left\{\begin{array}{l}D^1M^{cr}\\ D^1D^2M^{cr}\end{array}\right.$	**Triticum L., emend MacKey**		
		Dicoccoidea Flaksb.		
T. syriacum Bowden	$D^{cr}MS^1$	T. timopheevi Zhuk.	28	$AB(= AG)$
T. juvenale Thell.	DM^iC^u	T. turgidum (L.) Thell	28	AB
T. kotschyi (Boiss.) Bowden	C^uS^1			
T. ovatum (L.) Raspail	C^uM^o	Speltoidea Flaksb.		
T. triaristatum (Willd.) Godr. & Gren.	$\left\{\begin{array}{l}C^uM^t\\ C^uM^iM^u\end{array}\right.$	T. zhukovskyi Men. et Er.	42	AAB
		T. aestivum (L.) Thell	42	ABD
T. macrochaetum (Schuttl. & Huet. ex Duval-Jouve) Richter	C^uM^b	Triticale (Tscherm.) MacKey		
T. columnare (Zhuk.) Morris & Sears	C^uM^c	T. turgidosecale MacKey	42	ABR
		T. aestivosecale MacKey	56	$ABDR$
T. triunciale (L.) Raspail	C^uC	Trititrigia MacKey		
T. cylindricum Ces.	CD	T. turgidomedium MacKey	42	ABX
		T. aestivomedium MacKey	56	$ABDX$

Bowden (1959) pointed out that the morphological differences between *T. monococcum* L. and certain diploid species of *Aegilops* are barely adequate to separate the taxa at the specific level. He accepted the contemporary hypothesis that the genomes derived from the two diploid species of *Aegilops* had apparently undergone relatively little change since their incorporation into polyploid wheat. This meant that there were two nomenclaturally correct possibilities. The first is to place the parents of the polyploid wheats into two separate genera, a multispecific *Aegilops* and a monospecific *Crithodium* (*C. aegilopoides* Link = *Triticum monococcum*), and to put all the intergeneric allopolyploids into the genus × *Triticum*. The second possibility is to enlarge the genus *Triticum* to include both the wheat species and the *Aegilops* species. Bowden chose the latter course for reasons of simplicity of nomenclature, and because of the difficulties of species separation between the two genera at the diploid level. He retained the generic name *Triticum*, as it was the basis of the tribal name Triticeae, and recognized 22 species in three groupings (Table II).

Morris and Sears (1967), while they considered the method of classification proposed by Bowden (1959) to be satisfactory, modified it slightly by separating *T. timopheevi* from *T. turgidum* on the basis of genome differences and distributional data; they merged *T. turgidum* L. emend. var. *tumanianii* (Jakubz.) Bowden with the wild form, *T. timopheevi* (Zhuk.) Zhuk. var. *timopheevi*; and they merged *T. turcomanicum* (Rosh.) Bowden with *T. juvenale* Thellung on lack of evidence of any difference. They also added *T. columnare* (Zhuk.) Morris & Sears (Table III).

Recently, MacKey (1968) published a new classification founded on Bowden's alternative system (Table III). He explained that he chose this system on the basis of the continuing evolution of the polyploid *Triticum* in cultivation. He believed there was a possibility that a steady increase in the number of artificial allopolyploids produced from intergeneric hybridization and used in breeding would in time produce a discontinuity between the highly developed crop plant and both *Aegilops* and the primitive races of wheat, which would eventually be relegated to gene banks.

We have here then, two entirely different classification systems emphasizing different concepts of the genera concerned: Bowden's linked to the affinities between the wild species of *Aegilops* and the primitive diploid and tetraploid wheats; MacKey's emphasizing the differences between them. These opposing concepts have also resulted in a completely different nomenclature, thus setting the stage for a further period of confusion. Jakubziner's classification and nomenclature are being used by Harlan and de Wet (1971), Johnson and Dhaliwal (1976) and others; Bowden's by Sears (1969) and Feldman (1976). Before long, some workers in the field may begin to use MacKey's latest system also. In order to avoid the situation where three classifications are in concurrent use, it would be desirable for those studying wheat to agree as to which classification they prefer, or if this is not possible at the moment, at least to state in their publications which one they are using.

The wheats are an interesting case in that taxonomic evaluation has produced three distinct classifications in a short time. The cultivated pea, *Pisum sativum* L. subsp. *sativum* var. *sativum*, is a similar case. All recent biosystematic studies suggest that there are only two biological species (Waines, 1975), so the problem is less acute, mainly revolving around the question of whether var. *arvense* (L.) Poiret (being wild as well as cultivated, geographically widespread, and presumably antedating the cultivar group var. *sativum*), should be the specific name of the species containing the cultivar group (Blixt, 1972) or not (Davis, 1970). The degree of subdivision within the delimited species also varies (Table IV). Examples such as these are

TABLE IV. *The most commonly used classifications of* Pisum.

Boissier (1872)	Davis (1970)	Blixt (1972)
P. elatius Bieb.	*P. sativum* L.	*P. arvense* L., s.l.
	subsp. *elatius* (Bieb.) Aschers. & Graebn.	var. *abyssinicum* Braun
	var. *elatius*	
	var. *brevipedunculatum* Davis & Meikle	
P. humile Boiss.	var. *pumilio* Meikle	var. *humile* Boiss.
P. sativum L.	subsp. *sativum*	var. *sativum* L.
	var. *sativum*	
	var. *arvense* (L.) Poiret	var. *arvense* s.str.
P. fulvum Sibth. & Sm.	*P. fulvum* Sibth. & Sm.	var. *fulvum* Sibth. & Sm.

common; for instance, the same problem exists in the potato (Dodds, 1962; Hawkes, 1958), the tomato (Luckwill, 1943; MacBride, 1936), the chilli peppers (D'Arcy and Eshbaugh, 1974) and many others. This is not due to "vacillation and indecision among taxonomists" (Harlan and de Wet,1971), but to a real attempt by each taxonomist concerned to produce a classification which *in his considered opinion* approximates to the actual pattern of relationships. Even when there is a sound factual basis for a classification system, taxonomic opinions are going to differ, depending upon the outlook of the person concerned. When this situation occurs, it is the users of these classifications who must weigh the facts and ultimately decide which system gives the truest picture of their taxon.

THE CLASSIFICATION OF CULTIVARS

The most difficult problems are often met with at the infraspecific level, not only for the reasons given earlier, but additionally because it is possible to classify cultivars by any system, depending upon the purpose and the characters chosen, and because most classification at this level is done by workers not trained in taxonomic methods. The situation is not made easier by the fact that taxonomists have, in recent times, tended to ignore infraspecific variation in wild plants almost completely (Stace, 1976).

With cultivated plants there are three problems to be considered. Firstly, how to handle the large amount of conspicuous variation among closely related forms; secondly, how to differentiate between very similar entities at the cultivar level; and, lastly, what type of classification should be created to contain this information.

With regard to the first and third problems, many authors have attempted to devise classificatory systems into which they could fit the total range of variation found in different crop plants, applying an infraspecific nomenclature different from that used with wild plants. The difficulty here lies in finding a single flexible system which will satisfy all workers studying cultivars.

Recently, classifications have been proposed by Jirásek (1966), Jeffrey (1968) and Harlan and de Wet (1971) (Table V). Jirásek (1966), who produced the most detailed system, does not appear to be consistent. In a glossary of terms (Jirásek, 1961), he classified

TABLE V.   *Three recent suggestions for the classification of cultivated plants below the species level. It is not clear how "race" and "subrace" relate to earlier categories.*

Jirásek (1966)	Jeffrey (1968)	Harlan and De Wet (1971)
Specioid	Species (Subspecies)	Species
Subspecioid	Subspecioid	Subspecies
Cultiplex		
Subcultiplex		? ⎫
Convarietas	Convar	} Race
Subconvarietas	(Subconvar)	⎭
Provarietas	Provar	
Subprovarietas	(Subprovar)	? ⎫ Subrace
Conculta		⎬
Subconculta		⎭
Cultivar	Cultivar	Cultivar
Subcultivar		Line, Clone, Genotype

"conculta" as a synonym of "convarietas", but in his own classification, places it between "subprovar" and "cultivar", and also uses "convarietas" further up the hierarchy. If one of these items was removed, there would be nine categories below species level. On the other hand, Jeffrey (1968), using a similar system, reduced the number of infraspecific categories to four basic ranks, plus three more to be used as and when necessary. Harlan and de Wet (1971) proposed a system without any formal terminology. The only definition given is of "race", and they suggest that there should be a category below that of "cultivar." Jirásek (1966) also puts in a subcultivar category; however, he himself (1961) pointed out that according to the *International Code of Nomenclature for Cultivated Plants* (1958: art. 5):

> The term cultivar denotes an assemblage of cultivated individuals which are distinguished by any character (morphological, physiological, cytological, chemical, or others) significant for the purposes of agriculture, forestry, or horticulture, and which when reproduced (asexually or sexually), retain their distinguishing features.

The wording has not been changed. In the words of Jeffrey (1968), "it [the cultivar] is an atomic unit, further subdivision of which is not permitted."

There will always be some disagreement about the precise points of demarcation within a species, but this can be tolerated as long as the system is standardized to a reasonable extent. A suggested scheme, based on those given in Table V, is presented in Table VI. Between "cultivar" and "subspecioid" there is considerable flexibility, allowing classifiers to group their material according to its population structure under cultivation.

When the original wild-type of a cultivated plant has been identified, and relationships ascertained, wild and cultivated taxa should be distinguished within a united species as subspecies (feral) and "subspecioid" (cultivated). The "cultivar" has been fixed as the basic unit, and may be a genotype (clone), a pure line, or a closely selected outbreeding population. F_1 hybrids of crop plants reproduced by seed should be retained within the grouping "cultivar", and distinguished by the prefix "F_1 hybrid", as they do not retain their distinguishing features when their seeds are used to produce the next generation. The definition of "provar" is identical to that of "race" as defined by Anderson and Cutler (1942), and, like it, can be adapted to suit individual crops. In cereals such as maize or sorghum, the "provar" may well be the equivalent of a geographical race, whereas in many horticultural crops it could be a local group of genetically related inbred lines. The prefix "sub" may be used with multiconvar, convar and provar if necessary. The degree of subdivision depends on whether the group to be classified has been intensively studied, or only superficially surveyed.

TABLE VI. *A suggested system for use below the species level in cultivated plants, with category definitions which it is hoped are flexible enough to deal with most situations.*

Species	
Subspecioid	The equivalent of subspecies in wild plants.
Multiconvar	A group of related convars.
Convar	A group of related provars.
(Subconvar)	
Provar	A group of related cultivars with enough characteristics in common to permit their recognition as a group.
(Subprovar)	
Cultivar	see text

The classification of cultivars has proceeded rather slowly. In the cereals, and some other crop plants, the basis of an intraspecific classification was laid down during the last century by such workers as Alefeld (1866) and Körnicke (1885). These workers, and others, classified cultivated plants according to standard taxonomic practice, and gave all designated groups Latin names and formal descriptions, many of which are still in use today within species such as *Pisum sativum* (Lehmann, 1977) and *Hordeum sativum* Jessen (Grillot, 1959). Most modern cultivar classifications ignore previous work, however, and few people are prepared to make a detailed classification such as the one produced for *Lycopersicon esculentum* L. by Lehmann (1955). Furthermore, most modern classifications tend to use the particular part of the plant developed for human utilization for the major separation of cultivars into groups.

Daucus carota L. subsp. *sativus* (Hoffm.) Arc. appears to have a monospecific origin. Thellung (1926) suggested that it may have arisen as a result of hybridization between the wild carrots *D. carota* L. subsp. *carota* and *D. carota* L. subsp. *maximus* (Desf.) Ball. Apparently introduced into S.E. and S. Europe from the near east in the twelfth century, it was developed extensively in W. Europe, which became a secondary centre of variation between the sixteenth and nineteenth centuries (Banga, 1963). However, the main centre of variation is Afghanistan and E. Iran, although information is scanty. One of the earlier classifications was produced by Alefeld (1866), but the one presented is that of Banga (1964) (Table VII). The European and Afghanistan plants have been placed in separate "multiconvar" groupings for two reasons: a major difference in root colouring, and beyond that, a total lack of comparative research published on any other than the European groups of cultivars. As the carrot we know is a very highly selected crop plant, this lack of knowledge of the Eastern taxa is regrettable. This simple classification is natural only in that the earliest carrots introduced into W. Europe had very long roots, and development has been for shorter, more specialized forms.

Both genetics and total range of variation are far better known in *Pisum sativum* than in *Daucus carota*. The European cultivars of *P. sativum* L. subspd. *sativum* were very fully studied by Alefeld (1866) and Körnicke (1885). Their work, and further studies by Govorov (1928), are the basis of the very detailed classification system used by the germ-plasm bank at Gatersleben (Lehmann, 1977). This

TABLE VII. *A classification of European* Daucus *cultivars using the divisions given by Banga (1964).*

Multiconvar	Convar	Provar	Subprovar	Cultivar
		Daucus carota L. subspd. *sativus* (Hoffm.) Arcangeli		
I. *European* (Carotene carrot)	E. I (roots >25 cm)	"Field Carrots"		Long Orange Long Stump Winter, Long Orange Winter
	E. II (roots 20–25 cm)	"Conic-pointed"		St. Valery, Bauers Keiler Rote
		"Conic stump"		Flakkee, Flaro, Tendersweet, Imperator
		"Cylindric"		Long Red Coreless, Winter Perfection
	E. III (roots 15–20 cm)	"Conic-pointed" "Conic stump"		James' Intermediate Danvers, Feonia, Luc, Grosse Normande
	E. IV (roots 10–15 cm)	"Conic-pointed" "Conic stump"	Shoulders broad Shoulders medium narrow Shoulders narrow	Early Half Long Horn Guérande, Red Cored Chantenay Vertou, Montesson, Croissy Primerouge
		"Cylindric"	Shoulders medium narrow Shoulders narrow	"Nantes" group Amsterdam Forcing
	E. V (roots <10 cm)	"Cylindric"		"Early Short Horn" group, Davanture
		"Globular" "Flat-globular"		Grelot Parisienne
II. *Asiatic* (Anthocyanin carrot)	Range of variation not adequately documented			

classification contains five convars and 75 provars, each with a brief description. Work on pea cultivars did not really start in Britain until 1953 (Sneddon and Squibbs, 1958); it has continued on lines similar to recent W. European work (Fourmont, 1956; Kooistra, 1965). There has been no attempt to produce a natural as opposed to an artificial grouping of cultivars; the groups have not been given names and have been set out either in the form of dichotomous keys with cultivar descriptions (Sneddon and Squibbs, 1958), or simply as lists of cultivars under subheadings (Sneddon, 1970). There is also no consistency in the order in which characters have been grouped (Table VIII). This is rather disconcerting, considering that many countries are signatories of the *International Convention for the Protection of New Varieties of Plants,* under which plant breeder rights are agreed. An international classification system using agreed character orders for the identification of cultivars within each cultivated species would have considerable advantages.

An attempt to do this has been made for *Hordeum sativum* by

TABLE VIII. *A comparison of the characters used by different workers in the separation of cultivar groups in* Pisum sativum L. *subsp.* sativum *var.* sativum.

Kooistra (1965)	Sneddon and Squibbs (1958)	Sneddon (1970)
Flower White/coloured	*Pod* Normal/edible/purple	*Seed*
Pod wall Parchment/non	*Seed* Smooth/wrinkled	Smooth/wrinkled
Stature Bush/tall	*Cotyledons* Yellow/green	*Cotyledon*
Seed surface	*Pod shape* Pointed/blunt	Yellow/green
Smooth/wrinkled	*Stem* < 45 cm/45–75 cm/	*Pod apex*
Cotyledon colour	75–110 cm/ > 110 cm	Pointed/blunt
Green/yellow		
Combined with:		
Pod Blunt/pointed		
Seed colour at green shell		
stage Light-green/		
dark-green		
Earliness (*first fertile node*)		
Early/late		
Cultivar Groups		
64	36	8

Grillot (1959). He coordinated all the botanical systems of classification based on the main ear and spikelet characteristics. This produced 288 divisions, 104 of which corresponded to hypothetical cultivar groups not yet discovered or produced. This means that cultivars with new combinations of head characters can be included without distorting the framework of the system. Such an attempt deserves serious consideration, for it could also be applied to other crop plants.

In cultivated plants, as mentioned previously, a real problem has been that workers with a particular crop have used the portion of the plant selected for use as the major character in their classification. Where more than one species is cultivated showing parallel variation, as with the chilli peppers (Eshbaugh, 1975), true relationships can be obscured completely by concentrating on such a character. This approach has probably caused problems with crop plants of apparently monospecific origin also.

In the case of *Pisum sativum* cultivars, it could well be that use of less highly selected characters than the seed as primary groupings might produce a more efficient classification. Lamprecht (1956) postulated that in the evolution of the cultivated pea, selection of the recessive form (*a*) of the basic anthocyanin gene (*A*) was one of the early steps. He considered that other important changes were selection for a weak pod membrane (*V v*), wrinkled seeds (*R r*) and short internodes with a zig-zag stem (*Le le*). Davis (1970) used flower colour as one of the major divisions, separating var. *sativum* and var. *arvense*. One would logically continue using the flower characters of the standard, wing and keel; then the pod characters of wall structure, shape, end-form and size would be taken into account. Using the characters listed in Table IX, only after a large number of groups had been made would seed characters be considered, and after these, leaf and stem characters if necessary. This has not been attempted in the classification of *Pisum*, since workers have generally been mesmerized by the variation found in the seeds, and separated on these characters first. The system suggested would not be as continuous as present ones (there would be an unequal division at pod wall with/without parchment for instance), but it would operate in a similar manner to that of Grillot (1959) and hopefully it would be possible to separate out natural groupings within the species. No doubt a *Pisum* specialist could be more accurate as to character use

TABLE IX. *Suggested system of classification of* Pisum sativum *subspd.* sativum *convar.* sativum. *Only the flower characters venation and frilling would need defining, all other characters have been used previously.*

I. **Flower characters**
   Colour. White/cream
   Standard. Base—horizontal/rising/falling
       Apex—shallow/medium/deep
       Apex—mucronate/non-mucronate
       (?) Pattern of venation
       (?) Degree of frilling of margin
   Wing. Shape and position of insertion
   Keel. Shape and position of insertion

II. **Pod characters**
   Wall structure. With/without parchment
   Shape. Linear/curved
   End-form. Blunt/pointed
   Size. Large/small

III. **Seed characters**
   Form (with hilum central). Round/oval/drum/broad-cordate/
       cordate/long-cordate/square/flat-square
   Surface. Smooth/wrinkled
   Cotyledon colour. Green/yellow/orange
   Mature size. Large/medium/small/very small

IV. **Leaf characters**
   Tendrils. All/with leaflets/none
   Form. Wedge/oval
   Margin. Entire/toothed
   Colour. Yellow/green

V. **Stem characters**
   Habit. Bush/tall
   Internodes. Straight/zig-zag
   Form. Four-sided/fasciated
   First and second leaf scales. At least six types (Kooistra, 1965)

VI. **Maturity**
   Early/late

VII. **Cultivar**
   Semi-quantitative characters of seed testa, further shape characters, nodes to flower, further subdivision of pod characters, flower peduncle length, further subdivision of plant height.

and subdivision, but where there are well over 1000 cultivars to be identified (Kooistra, 1965), the principle that characters which do not give good separation should be relegated, must be strictly applied.

Similar problems to the above occur within other cultivated species. In *Phaseolus vulgaris* L., Chopinet *et al.* (1950) and North and Squibbs (1952) used seed colour, shape and size as the most important characters for identification. However, with the recent increase in numbers of cultivars with white seeds, it has been necessary to place much more emphasis not only on pod characters in particular, but also on leaf and stem characters, to give satisfactory cultivar separation within this one grouping (Sneddon, 1969). It is considered that a classification similar to that suggested for *Pisum* would be more useful. The situation in *Lactuca sativa* L. is rather different in that the classification produced by Rodenburg and Huyskes (1965) is aimed at identification of cultivars in the seedling stage, whereas those of Rodenburg (1960) and Bowring (1969) are concerned more with adult plant characteristics. It should also be possible to construct a comprehensive classification for this taxon by combining both the above sets of characteristics.

Any good classification will provide a description of each provar, stressing the internal similarities and the differences between it and other provars. For instance, Sneddon (1969) has shown that the white-seeded group of *Phaseolus vulgaris* L. which he studied is a convar with two provars, for which he gives brief descriptions (Table X). Within each provar are subprovars, each with a description, and finally each cultivar is described, although more briefly than those studied by North and Squibbs (1952). There is a general trend towards discarding cultivar descriptions, which in the opinion of the writer is a retrograde step; however, according to Sneddon (personal communication), "a generally held view is that formal descriptions have limited value and can be misleading". Properly written cultivar descriptions with details of parentage, and, where it varies, chromosome number, are neither of these things. Publication is important both for the study of the evolution of the cultivar group, and also for providing a record of the range of variation present at any one time or place, and of the size of the gene pool. Furthermore, the lack of adequate descriptions for cultivars causes endless confusion, and encourages the giving of new names to cultivars which differ very little from their predecessors. This causes unnecessary work for those reg-

TABLE X. *Classification of* Phaseolus vulgaris *L. subspd.* vulgaris *convar. nanus (Jusl.) Aschers., white-seeded provar. Modified from Sneddon (1969).*

Provar A. Pods medium-long; round or eight-shaped in section. Beaks medium or long, inconspicuous constrictions between seeds. Bush much branched, erect; leaves profuse, mainly acute or rounded at base.

	Subprovars						
	1	2	3	4	5	6	7
Pods medium or pale green	×	×					
Pods dark green			×	×		×	×
Pods medium-dark grey-green					×		
Pods pointed	×			×	×		×
Pods semi-blunt		×	×	×	×	×	×
Pods blunt			×			×	
Pods rough textured						×	×
Pods smooth	×	×		×	×		
Neither rough nor smooth			×	×			×
Pods shiny				×			
Pods matt	×	×	×			×	×
Neither shiny nor matt				×	×	×	×
Pods with coarse beaks				×			
Pods without coarse beaks	×	×	×		×	×	×

Provar B. Pods short-medium length; oval-round or round in section. Beaks mainly short and hooked or long-curved; mainly constricted between seeds. Bush slightly or medium branched, low, spreading, or medium erect; leaves sparse or medium dense, mainly with obtuse base and non-acuminate apex.

	Subprovars			
	8	9	10	11
Pods relatively constricted	×	×	×	
Pods relatively unconstricted				×
Pods medium or pale green	×	×	×	
Pods dark green			×	×
Pods pointed			×	×
Pods semi-blunt	×	×		×
Pods blunt	×	×		
Beak long and curved		×		
Beak not long and curved	×	×		×
Bush low or spreading	×			
Bush medium erect		×	×	×

istering and testing these cultivars under plant breeders' rights Acts. This problem of inadequate descriptions was commented on by Hatton (1920), but the situation has not improved over the last 57 years.

CONCLUSIONS

The above discussion illustrates that a great deal of the work on the wild relatives of cultivated plants, on the variation found within these wild taxa, and also studies on the phylogenetic relationships between related wild and cultivated groups, is left to plant breeders and others. For this reason, classification work is neglected, although it is an essential preliminary to any serious research in the field of plant breeding, especially in those crops where germ-plasm from wild species is used to any extent. It is not uncommon for plant breeding establishments to have neither herbaria for storing specimens of the wild species and named cultivars used in their research, nor even a collection of photographic vouchers of the type pioneered by Anderson. It is also uncommon for workers in this field to give authorities for the species names they use in their publications. In hybridization work in particular, if no voucher specimens are preserved, previous work can be invalidated or made incapable of confirmation.

Natural variation within the limits of the species or subspecies is not often classified in wild plants. However, in cultivated plants, given the cytogenetic, genetic, distributional, morphological and biochemical data available, it should be possible to produce natural groupings of cultivars which reflect the whole pattern of variation within one species or species group. This will create an efficient cultivar identification system which, where the wild progenitors are known, may perhaps make it possible to relate different cultivar groups to different nodes of variation in wild populations. This work cannot be undertaken in isolation—it needs to be done in conjunction with studies on the genetics, breeding and origins of cultivated plants. Such a step is necessary to give cultivar classification work its true place as an important and integral part of the study of plant variation in general, and to encourage taxonomists with some interest in crop plants to accept the challenge of a field which is in serious need of their services.

REFERENCES

Alefeld, F. (1866). "Landwirthschaftliche Flora." Weigandt and Hempel, Berlin.

Anderson, E. (1941). The technique and use of mass collections in plant taxonomy. *Ann. Mo. bot. Gdn* **28**, 287–292.

Anderson, E. (1947). Field studies of Guatemalan maize. *Ann. Mo. bot. Gdn* **34**, 433–467.

Anderson, E. (1954). "Plants, Man, and Life", pp. 92–95. A. Melrose, London.

Anderson, E. and Cutler, H. C. (1942). Races of *Zea mais* L. I. Their recognition and classification. *Ann. Mo. bot. Gdn* **29**, 69–88.

Banga, O. (1963). "Main Types of the Western Carotene Carrot and their Origin." Tjeenk Willink, Zwolle.

Banga, O. (1964). Identification of western orange carrot varieties (*Daucus carota* L.) *Proc. Int. Seed Test. Ass.* **29**, 957–961.

Blixt, S. (1972). Mutation genetics in *Pisum* L. *Agri. Hort. Genet.* **30**, 1–293.

Boissier, E. (1867–1888). "Flora Orientalis." University of Geneva, Geneva.

Bowden, W. M. (1959). The taxonomy and nomenclature of the wheats, barleys, and ryes, and their wild relatives. *Can. J. Bot.* **37**, 657–684.

Bowring, J. D. C. (1969). The identification of varieties of lettuce (*Lactuca sativa* L.). *J. natn. Inst. agric. Bot.* **11**, 476–498.

Chopinet, R., Trebuchet, G. and Drouzy, J. (1948–1950). Essai de classification et d'identification des principales variétés de haricots cultivées en France. *Rev. hort., Paris* **120–122**.

D'Arcy, W. G. and Eshbaugh, W. H. (1974). New World peppers (*Capsicum–Solanaceae*) north of Columbia: A résumé. *Baileya* **19**, 93–105.

Davis, P. H. (1965 *et seq.*). "Flora of Turkey and the East Aegean Islands." University Press, Edinburgh.

Davis, P. H. (1970). *Pisum* L. *In* "Flora of Turkey and the East Aegean Islands" (P. H. Davis, ed.), pp. 370–372. University Press, Edinburgh.

Dodds, K. S. (1962). Classification of cultivated potatoes. *In* "The Potato and its Wild Relatives". (D. S. Correll, ed.), pp. 517–539. Texas Research Foundation, Renner.

Eig, A. (1929). Monographische — Kritische Uebersicht der Gattung *Aegilops. Repert. Spec. Nov. Reg. Veget. Beih.* **55**, 1–228.

Eshbaugh, W. H. (1975). Genetic and biochemical systematic studies of chili peppers (*Capsicum–Solanaceae*) *Bull. Torrey bot. Club* **102**, 396–403.

Exell, A. W.and Wild, H. eds. (1960 *et seq.*) "Flora Zambesiaca." Crown Agents, London.

Feldman, M. (1976). Wheats. *In* "Evolution of Crop Plants" (N. S. Simmonds, ed.), pp. 120–128. Longman, London and New York.

Fourmont, R. (1956). "Les Variétés de Pois Cultivées en France." Inst. Nat. Rech. Agron., Paris.

Goodspeed, T. H. (1954). "The Genus *Nicotiana*". Waltham, Massachusetts.

Govorov, L. I. (1928). The peas of Afghanistan (in Russian). *Trudi po Prikl. Bot.* **19**.

Grillot, G. (1959). La classification des Orges cultivées (*Hordeum sativum* Jessen). *Ann. d'Amelior. des Plantes* **9**, 445–552.

Hackel, E. (1887). *Triticum* L. In "Die Natürlichen Pflanzenfamilien" (H. G. A. Engler and K. A. E. Prantl, eds), p. 80. ii. 2. W. Engelmann, Leipzig.

Harlan, J. R. and de Wet, J. M. J. (1971). Toward a rational classification of cultivated plants. *Taxon* **20**, 509–517.

Hatton, R. G. (1920). Blackcurrant varieties—a method of classification. *J. Pomol.* **1**, 65–80, 145–154.

Hawkes, J. G. (1958). Kartoffel. I. Taxonomy, cytology, and crossability. In "Handbuch der Pflanzenzuchtung" (W. Rudorff and T. E. M. Roemer, eds), 2nd edition., Vol. 3, pp. 1–43. P. Parey, Berlin.

Heiser, C. B. Jr. (1968). Systematics and the origin of cultivated plants. *Taxon* **13**, 36–45.

Howard, R. A. *et al.* eds. (1968 *et seq.*) "Flora Neotropica." Hafner, New York.

Jakubziner, M. M. (1958). New wheat species. *Proc. 1st Int. Wheat Genet. Symp.* (B. C. Jenkins, ed.), pp. 207–220. University of Manitoba, Winnipeg.

Jeffrey, C. (1968). Systematic categories for cultivated plants. *Taxon* **17**, 109–240.

Jirásek, V. (1961). Evolution of the proposals of taxonomical categories for the classification of cultivated plants. *Taxon* **10**, 34–45.

Jirásek, V. (1966). Systematica Kulturnich rostlin a jeji tridici kategorie. *Preslia* **38**, 267–284.

Johnson, B. L. and Dhaliwal, H. S. (1976). Reproductive isolation of *Triticum boeoticum* and *T. urartu*, and the origin of the tetraploid wheats. *Am. J. Bot.* **63**, 1088–1094.

Kihara, H. (1954). Considerations on the evolution and distribution of *Aegilops* species based on the analyser-method. *Cytologia* **19**, 336–357.

Knight, R. L. and Keep, E. (1964). Soft fruit breeding. *Ann. Rep. E. Malling Res. Stn.* **51**, 158–160.

Körnicke, F. (1885). Die Arten und Varietaten des Getreides. In "Handbuch des Getreidebaues" (F. Körnicke and H. Werner, eds.), Vol. I. P. Parey, Berlin.

Kooistra, E. (1965). Identification research on pulses. *Proc. Int. Seed Ass.* **29**, 937–948.

Lamprecht, H. (1956). *Pisum sativum* L. oder *P. arvense* L., Eine nomenklaturische studie auf genetischer basis. *Agri. Hort. Genet.* **14**, 1–4.

Larsen, J. (1974). The role of chromosome interchanges in the evolution of hexaploid wheat (*T. aestivum*). *Proc. 4th Int. Wheat Genet. Symp.* (E. R. and L. M. S. Sears, eds), pp. 87–93. Columbia University, Columbia.

Lehmann, Chr. O. (1955). Das morphologische system der Kulturtomaten (*Lycopersicon esculentum* Mill.). *Der Züchter* **3**, 64.

Lehmann, Chr. O. (1977). *Pisum. In* "Index Seminum", pp. 171–178. Zentralinstitut für Genetik und Kulturpflanzenforschung, Gatersleben.

Luckwill, L. C. (1943). The evolution of the cultivated tomato. *J. R. hort. Soc.* **68**, 19–25.

Lundquist, K. (1960). On the origin of cultivated lettuce. *Hereditas* **46**, 319–349.

MacBride, J. F. (1936 *et. seq.*). "The Flora of Peru" Field Museum of Natural History, Chicago.

MacKey, J. (1968). Relationships in the Triticinae. *Proc. 3rd Int. Wheat Genet. Symp.* (K. W. Finlay and K. W. Shepherd, eds.), pp. 39–50. Butterworth, Sydney.

Maesen van der, L. J. G. (1972). *Cicer* L., a monograph of the genus with special reference to the chickpea (*C. arietinum* L.), its ecology and cultivation. *Meded. LandbHoogesch.* **72/10**.

Martius von, C. F. Ph., ed. (1840–1906) "Flora Brasiliensis." Facs. Reprint (1965). J. Cramer, Weinheim.

Morris, R. and Sears, E. R. (1967). The cytogenetics of wheat and its relatives. *In* "Wheat and Wheat Improvement" (K. S. Quisenberry and L. P. Reitz, eds), pp. 19–87. Am. Soc. Agron., Madison.

Nordberg, G. (1967). The genus *Sanguisorba* section Poterium. Experimental studies and taxonomy. *Op. bot. Soc. bot. Lund.* **16**, 1–166.

North, C. and Squibbs, F. L. (1952). A description of dwarf French Bean varieties grown in the U.K. *J. natn. Inst. agric. Bot.* **6**, 196–211.

Ornduff, R. (1966). A biosystematic survey of the goldfield genus *Lasthenia*. *Univ. Calif. Pub. Bot.* **40**, 1–92.

Percival, J. (1921). "The Wheat Plant, a Monograph." Duckworth, London.

Persson, K. (1974). Biosystematic studies in the *Artemisia maritima* complex in Europe. *Op. bot. Soc. bot. Lund.* **35**, 1–188.

Rechinger, K. H. (1963 *et seq.*) "Flora Iranica." Akademische Druck-u. Verlagsanstalt, Graz.

Riley, R. (1965). Cytogenetics and evolution of wheat. *In* "Essays on Crop Plant Evolution" (J. B. Hutchinson, ed.), pp. 103–122. Cambridge University Press, Cambridge.

Riley, R. and Chapman, V. (1958). Genetic control of the cytologically diploid behaviour of hexaploid wheat. *Nature, Lond.* **182**, 713–715.

Rodenburg, C. M. (1960). "Varieties of Lettuce; an International Monograph." Tjeenk Willink, Zwolle.

Rodenburg, C. M. and Huyskes, J. A. (1965). The identification of varieties of lettuce, spinach, and whitloof chicory. *Proc. Int. Seed Test. Ass.* **29**, 963–976.

Rubenstein, J. M. and Sallee, P. J. (1974). The genomic relations of

Triticum kotschyi to the 'S.' genome diploids. *Proc. 4th Int. Wheat Genet. Symp.* (E. R. and L. M. S. Sears, eds.), pp. 95–100. Columbia University, Columbia.

Sauer, J. D. (1950). The grain amaranths: a survey of their history and classification. *Ann. Mo. Bot. Gdn.* **37**, 561–632.

Sears, E. R. (1969). Wheat cytogenetics. *Ann. Rev. Genet.* **3**, 451–468.

Shands, H. and Kimber, G. (1974). Reallocation of the genomes of *Triticum timopheevi* Zhuk. *Proc. 4th Int. Wheat Genet. Symp.* (E. R. and L. M. S. Sears, eds.), pp. 101–108. Columbia University, Columbia.

Sneddon, J. L. (1969). Identification features of white seeded stringless varieties of French Beans (*Phaseolus vulgaris* L.) *J. natn. Inst. agric. Bot.* **11**, 476–498.

Sneddon, J. L. (1970). Identification of garden pea varieties (1) Grouping arrangement, and use of continuous characters. *J. natn. Inst. agric. Bot.* **12**, 1–16.

Sneddon, J. L. and Squibbs, F. L. (1958). Classification of garden pea varieties. *J. natn. Inst. agric. Bot.* **8**, 378–422.

Stace, C. A. (1976). The study of infraspecific variation. *Curr. Adv. Plant Sci.* **23**, 513–532.

Stuart, D. C. (1974). Some (taxonomic) problems at the cultivar level. *Taxon* **23**, 179–184.

Thellung, A. (1926). *Daucus. In* "Illustrierte Flora von MittelEuropa" (G. Hegi, ed.), Vol. 5, pp. 1501–1526. Lehmann, München.

Turrill, W. B. and Milne-Redhead, E., eds. (1952 *et seq.*). "Flora of Tropical East Africa." Crown Agents, London.

Ugent, D. (1968). The potato in Mexico: Geography and primitive culture. *Econ. Bot.* **22**, 109–123.

Vardi, A. (1974). Introgression between different ploidy levels in the wheat group. *Proc. 4th. Int. Wheat Genet. Symp.* (E. R. and L. M. S. Sears, eds.), pp. 131–141. Columbia University, Columbia.

Waines, J. G. (1975). The biosystematics and domestication of peas. *Bull. Torrey bot. Club* **102**, 385–395.

Zohary, D. and Feldman, M. (1961). Hybridisation between amphidiploids and the evolution of polyploids in wheat (*Aegilops–Triticum*) group. *Evolution* **16**, 44–61.

Chapter 7

The Taxonomist's Role in the Conservation of Genetic Diversity

J. G. HAWKES

THE BASIS OF DIVERSITY

Perhaps the most important advance in evolutionary taxonomy during the last 40 to 50 years has been the clear demonstration of genetic diversity within the commonly accepted concepts of the species. It is now hardly necessary to apologize for using the word "concepts" rather than "concept", since biosystematic and genetical studies have made it abundantly clear that no one definition of a species can be rigidly adhered to. Furthermore, students of evolution are well aware of the genetic variability occurring within all living organisms through the processes of mutation, recombination and selection.

What is not quite so clear is how much of this infraspecific diversity should receive formal taxonomic recognition. To botanists of the period up to the 1920s or 1930s the problem was simple. Any distinguishable morphological variant was to be given status as a *forma* at least and very often even a higher taxonomic category. It is now agreed, however, that most of such minor categories are based either on some particular environmental effect which does not presuppose a genetical determinant and is hence inconstant, or on the effect of a single gene or a group of closely linked genes which are of insufficient frequency to warrant more than passing mention.

Apart from these variants we also know of many more, such as the adaptive gene complexes with phenotypic expression that occur in certain areas of the distribution of a species. These may be given subspecies rank or even classed as distinct species if the boundaries between them are sufficiently clear cut. Under these circumstances it is assumed that the series of populations in such areas have achieved at least some degree of reproductive isolation.

Then again, differences in chromosome number and behaviour may also, though not invariably, be correlated with clear morphological differences, so that taxonomic status may be justifiable and indeed useful.

However, for each adaptive trait or adaptive complex for which taxonomic status at or below the species level can be justified, there must be many thousands where it cannot, through lack of the linked morphological features which all taxonomists must have recourse to in the long run, even though cytogenetical, biochemical and physiological characters can be of very great use in helping to distinguish and define taxa.

Thus the study of variation within and between populations and the linking of this where appropriate to clearly marked morphological features so as to define taxa, often at the infraspecific level, is a very appropriate activity for the modern experimental taxonomist. In this respect he is moving close to the working area of the population geneticist. Each is interested in the study of variation, but from slightly different view-points. Whereas the population geneticist is concerned with the processes by which genetic variation is engendered, maintained or changed within and between populations, the taxonomist is more interested in the variation as such, and the definition of taxonomic categories based on it, where he feels them to be appropriate.

Thus, the taxonomist has developed a trained eye for morphological differences. If he is a chemotaxonomist he will be clearly interested in variation patterns of proteins, amino acids, isoenzymes, essential oils, flavonoids, terpenes, anthocyanins and many other classes of compounds. The pure biochemist is also concerned with these, but, whereas the biochemist's interests lie primarily with the chemicals themselves and the pathways by which they are engendered, the chemotaxonomist is more concerned with the whole organism and the comparison of the chemical spectra within the taxa he is studying.

In the foregoing paragraphs no mention has been made of ecologically adapted diversity within and between species. Clearly, individuals and populations must be adapted to their environment if they are to survive, and the biosystematist is well aware of the presence of adaptive variation within populations of the taxa he is studying. Again, adaptation does not need to be defined in taxonomic terms; most adaptive traits are highly complex and of a physiological nature, with little or no accompanying morphological differences. However, this is not always so, and every student of experimental taxonomy is well aware of the distinctive morphological features in high-mountain and maritime ecotypes or subspecies of many species in the European flora. So here again, some ecologically adapted variants are given taxonomic status because of the clear morphological differences linked to physiological differences; but where such morphological differences do not occur there is no point in trying to recognize taxa based on features only discernible in the laboratory.

The identification of clinal variation, as many workers have pointed out, linked generally to differences of habitat or climate, can rarely lead to the recognition of formal taxonomic groupings because of the lack of clear discontinuities. Nevertheless, clines exhibit genetical variation and their presence must be recognized and studied by taxonomists.

From the point of view of the conservation of genetic resources, one lesson for the taxonomist is clear. He tends on the whole to pay more attention to morphological variation since this is still the basis for the definition and recognition of taxa. Anatomical, cytological, genetical, biochemical and physiological data are all to a greater or lesser degree used to deepen and clarify our knowledge of taxonomic relationships where correlation is shown with morphological characters. However, we tend to discard as of lesser importance variation which does not correlate well, since this is part of our training. This is a danger that we must be aware of. The reasons for such a warning will become apparent later in this chapter, but it must be emphasized at once that all types of variation may be of potential importance and should thus not be overlooked in genetic resources conservation.

THE CONSERVATION OF GENETIC DIVERSITY

The conservation of tracts of wild vegetation in its natural or semi-natural state has captured the imagination of biologists and the general public in these days of ever mounting pressure for change and development stemming from the spread of urbanization and of modern high-technology farming. We appreciate the comforts of modern living but at the same time we need to renew our spirit by contact with raw or only slightly tamed nature, and we are thus anxious to conserve as much natural vegetation as we can.

For those of us living in highly industrialized countries this is generally true. In the Third World, however, the point of view is understandably rather different. A reasonably decent standard of living is not within the grasp of more than a small minority; hence the emphasis is on development and less on conservation. A very practical approach is adopted of being in favour of conservation when it is of value now or in the foreseeable future but not otherwise. Hence, the conservation of vast areas of forest which could, if felled, provide a good source of foreign exchange from the export of its timber, is not so enthusiastically agreed to. On the other hand, one must, to be fair, give credit to those far-sighted politicians and administrators in Third World countries who see the need for conservation of rare and threatened species and ecosystems, and are attempting to carry out plans for biosphere reserves in collaboration with international agencies such as UNESCO and FAO.

The International Union for the Conservation of Nature and Natural Resources (IUCN) is of course well aware of the need to conserve, not only special areas of ecological and natural resource importance but also actual species that are rare or under threat of complete extinction. This is an activity that is highly deserving of respect and support. It is one, moreover, where the taxonomist, together with the ecologist, must play a central role. A knowledge of the floras of countries and regions, and of the distribution of taxa within these areas, as well as an understanding of their adaptation to certain zones, is essential to conservation work of this kind, and in most parts of the world only the taxonomist possesses such knowledge.

The IUCN Threatened Plants Committee and the *Red Data Book for Threatened Plant Species* started by R. Melville at Kew (see Lucas, 1976) is an entirely praiseworthy attempt to conserve on a world scale plant species that are in danger of extinction. The Convention on International Trade in Endangered Species of Wild Fauna and Flora (Simmons *et al.*, 1976) is also an essential weapon in the task of preventing the total extinction of taxa of interest and potential value. In this chapter, however, my task is to draw attention to an equally important type of conservation which is closely linked to the ones described above, and which can in no way be considered as opposed to them. I refer to the conservation of genetic variation within species, both wild and cultivated.

The problem was first defined by applied scientists, chiefly in the fields of agricultural and plant breeding research, who were anxious to ensure the preservation of as wide a range as possible of genetic variability to form a continued basic supply for breeding new cultivars—hence the phrase "genetic resources". One must, in fact, return to the work of the Russian geneticist, N. I. Vavilov, who in the 1920s and 1930s showed the rest of the world the tremendous genetic variation of our ancient crop plants in a certain limited number of world centres of diversity. These he classed as centres of origin of crop species and drew attention to the useful economic features of adaptation and disease resistance that could be found in them. As a geneticist of great breadth of vision he recognized the importance of taxonomic studies in cultivated plants, as few have done before or since, in helping to characterize and define the bewildering variation which he and his colleagues saw and collected as living samples in most parts of the world (see Vavilov, 1940). He clearly showed the need for biosystematic approaches in a study of cultivated plant taxonomy and pointed out the evolutionary aspects of such work in our ancient crop plants, which have evolved under domestication from wild prototypes (which mostly still exist) over the last 10,000 years. After his death in the early 1940s work along these lines has continued, using the newer techniques of numerical chemo-taxonomy which were devised chiefly for the study of wild taxa. Data from archaeology have given to evolutionary studies on cultivated plants a most valuable fourth or time dimension.

The conservation of crop plant genetic diversity

In Vavilov's day genetic resources of cultivated plants and related wild species were being studied in their centres of genetic variability—the so-called "gene centres"—and were also being used as source material for plant breeding, though perhaps in most countries only to a limited extent.

In the 1950s and 1960s, the spread of new cultivars and the adoption of new farming techniques in the Third World countries where the gene centres mostly occurred, were leading to disastrous consequences. By a curious paradox the cultivars, which had been bred from materials stemming from the gene centres, were themselves taking the place of these materials and thus causing their destruction. A process of "genetic erosion" had begun which has gathered momentum through the past two decades to the extent that nothing is likely to be left by the end of this century in what are already becoming known as "former centres of diversity". This process clearly cannot be allowed to continue, since the development of cultivars with higher yields and disease resistance, and better ranges of environmental adaptation, are essential to the survival of mankind, and the gene centre material is an essential basis for their production.

Much has been written about the conservation of genetic resources of crop plants. The United Nations Food and Agriculture Organization (FAO) with the help of the International Biological Programme (IBP) organized three technical conferences on the problem, from which two books on this subject have been published (Frankel and Bennett, 1970; Frankel and Hawkes, 1975). Both deal with problems of variation in cultivated plants and are thus of interest to taxonomists. The first also looks at the specific problem of the taxonomy of cultivated plants (Baker, 1970; Hawkes, 1970) and gives a number of crop examples in relation to exploration work.

Much research related to the exploration, conservation, evaluation and utilization of crop genetic resources lies outside the field of taxonomy. Long-term conservation of seeds in gene banks, for instance, is the concern of the seed physiologists (Harrington, 1970; Roberts, 1975; Villiers, 1975; Sakai and Noshiro, 1975). Conservation of meristem and cell cultures lies within the field of tissue culture specialization (Morel, 1975; D'Amato, 1975; Henshaw,

1975). Documentation and data management is yet another specialist field (see Rogers *et al.*, 1975; Hersh and Rogers, 1975). Nevertheless, taxonomists are becoming well aware of the value of electronic data processing in taxonomy (see Brenan *et al.*, 1975; Simmons *et al.*, 1976: Section IV, Documentation). Evaluation of materials for breeding ranges through a wide field of specialities from phytopathology and physiology to biochemistry (see Chapters 16–20 in Frankel and Hawkes, 1975).

In the tasks of exploration and sampling we are nearer home. Taxonomists are becoming more and more aware of the need to sample populations rather than being content to make a seed collection from a single plant or to sample a vegetatively propagated taxon from a single point. Hence, advice from the population geneticist is obviously essential in working out sampling strategies so that the taxonomist can be certain of capturing the highest possible proportion of the genetic variation within a species (see Hawkes, 1976).

The documentation of samples is of obvious importance, and is an area where the taxonomist can play a major role through his experience as a collector and his need for adequate collection data when checking and comparing herbarium specimens (Gomez-Pompa *et al.*, 1975). In the past, those geneticists and plant breeders who have made living collections for genetic resources purposes have either erred too far in one direction and have been content with a note of the country and nearest city to the collection site, or have advocated the noting down of so much information that it is doubtful whether they would have much time left to collect actual plant samples (see, for example, pp. 157–179 in Bennett, 1970; Godron and Poissonet, 1970). Attempts are now being made to provide minimum collecting records for cultivated plants to which very few collectors would take exception and which might also be suitable for wild plants (pp. 150–153 in Hawkes, 1976).

Concerning the exploration and identification of crop-plant genetic resources, taxonomists have not perhaps played a prominent role, having left this to the geneticists and agricultural scientists, few of whom have been trained as professional taxonomists. The works of Harlan (Harlan, 1966; Harlan and de Wet, 1971, 1972), Zohary (1970, 1971, 1972), de Wet (de Wet *et al.*, 1976) and several others are notable exceptions. The present writer has stressed the need for experimental taxonomists to enter this field (see Hawkes, 1970) since they possess a knowledge of the techniques necessary for studying the taxonomy and evolutionary relationships of crop plants. These are taxa for which ancestral forms generally still exist, and even preserved archaeological material can be used to confirm or otherwise their postulated course of evolution under domestication.

The conservation of timber and fruit tree genetic diversity

The basic problems of conservation of timber and fruit tree genetic diversity are similar to those of crops but for a variety of reasons the solutions are not so easy to arrive at. Whereas crop plants are for the most part annuals, biennials or short-lived perennials which can be stored in seed banks, generally under reduced temperature and moisture (Roberts, 1975) for long periods (10–50 years or even longer) before renewal by taking them through a single life cycle ("regeneration"), trees are normally very long-lived and the seed cannot be regenerated in a single season, as that of crop plants can. Problems of seed sampling are more difficult, through the sheer size of the trees, the very irregular seedling cycles and, in tropical forests, the isolation of individuals one from the other. Random sampling, which is a *sine qua non* for genetic resources purposes and which can be undertaken without trouble in cereals, presents a tremendous problem with forest and fruit trees. We hardly know the elementary outlines of the reproductive biology of tropical tree species, let alone what really constitutes a "population".

As though this were not enough, the seeds of many forest trees and nearly all fruit trees cannot be stored by the normal methods of low temperature ($-18°C$) and reduced humidity ($\pm6\%$). Such seeds have been termed "recalcitrant" for obvious reasons (Roberts, 1975), and research is urgently needed to formulate methods for conserving them. Alternative ways of conservation must be used, such as the designation of reserve areas where the natural genetic diversity, together with the total ecosystems, can be preserved (see Sykes, 1975). This method can be alternated with the planting of fruit tree or forest tree nurseries.

A grave disadvantage of both these methods is that neither can hope to solve the problem of conserving anything approaching the total genetic diversity of the species in question. Reserves can be most valuable in conserving that part of the variation which occurs in them but cannot hope to coincide with the total distribution area of all but restricted endemics or taxa of great rarity. Even so, where tropical forests are in danger of being destroyed completely, forest reserves are clearly of the greatest importance. Forest and fruit tree nurseries, again, for reasons of size, can only hope to preserve a small part of the genetic diversity, but this is better than nothing, if conservation in the form of seeds is not possible.

Within this field of endeavour taxonomists, together with ecologists, are playing a most important role (see Burley and Styles, 1976) in studying the reproductive biology of tropical trees together with various strategies of conservation. Work in south-east Asia in this general field has also been discussed by Sastrapradja (1975), Guzman (1975), Whitmore (1975a), Lamoureux (1975) and others in a book recently edited by Williams *et al.* (1975) resulting from a conference on *South-east Asian Plant Genetic Resources*, held in Bogor, Indonesia earlier that year.

In the tasks of designating forest reserves and in studying the full economic potential of timber trees, oil nuts, gums, resins, rubber trees, rattans, bamboos and fibres (see Lamoureux, 1975), taxonomists are urgently needed. Too many of us have trodden the well worn paths of temperate plant taxonomy and have encouraged our research students to do likewise. As Whitmore (1975b) says:

> Forests are a biochemical storehouse. In the longer prospect they can be expected to play an increasing part in maintaining the world's civilizations.

Until we have studied these storehouses taxonomically, ecologically and biochemically, we shall not be in a position to influence the decisions of administrators and politicians, with the consequence that the storehouses may be destroyed before even a primary taxonomic survey has been made, let alone a more detailed collection of their genetic riches.

The conservation of genetic diversity in wild herbaceous species

This theme has been even less discussed than the previous one. We know a reasonable amount about those wild species that are related to our major crops, since they have been collected for evaluation in respect of disease resistance that may be transferred to the crops themselves. In some, such as potatoes and wheat, the success has been fairly good; in others, such as maize, the problems of species "incongruity" are not inconsiderable (as, for example in crossing *Zea* and *Tripsacum*). Nevertheless, breeders with nearly all crops are looking for sources of genetic resistance to fungal, viral, bacterial, insect and nematode parasites in primitive cultivated forms and in related wild species. The prospects are promising but, just as we lack good basic work on the experimental and evolutionary taxonomy of cultivated species, so we also lack this in respect of their wild prototypes and relatives. Here is a field in which all too few taxonomists have dared to venture, yet the results when they come are rewarding. Taxonomists should be in a position to elucidate relationships between wild and cultivated plants so that breeders may obtain an idea not only of their potential in transferring characters of value, but simply to know what is there. A good, flexible taxonomic system with broadly conceived species and a minimum of name changes will help breeders to understand and use the available material and at the same time develop a healthy respect for taxonomy as a modern experimental discipline.

Most gene banks of crop plants possess very few wild species accessions (apart from genera such as *Aegilops*, *Lycopersicon*, *Solanum*, *Avena* and some others). A really wide genetic base of wild material is even rarer in gene banks. Yet a wide genetic base is essential, since many instances are known of a valuable gene conferring disease resistance, which occurs only in an extremely restricted part of the total range of a species. This makes it essential to sample widely over the whole species range and shows the danger of relying on reserve areas which necessarily must be restricted to only a minor part of the total range.

We come now to those wild species that are not closely related to crop plants. These in fact constitute the majority of plant species throughout the world. The task of making adequate samples of these would be prodigious, and in fact impossible. How then should we devise a scheme of priorities, so that at least some, and perhaps those of greatest potential value, could be preserved? There is only one gene bank in existence at present which devotes itself wholly to the conservation of genetic resources of wild species. This is the seed bank at Wakehurst Place, the satellite garden of the Royal Botanic Gardens, Kew (see Thompson, 1970, 1974, 1975, 1976). The conserved seeds of crucifers by Gomez-Campo (1972) from the west Mediterranean region is of interest and value, and could well be extended for other parts of the Mediterranean basin and for other families. In Birmingham we ourselves keep a large living collection of Solanaceae from all parts of the world, though this cannot be claimed at present to be anywhere near completion. Systematic collections to sample a wide diversity within each species are still lacking, though certain genera receive special treatment. Other smaller collections exist but for various reasons could hardly be classed as gene banks, in which certain minimal standards of collection, storage and regeneration are required.

Returning to the question of what to put into a bank of this sort, one is faced with numerous categories of choice, some often conflicting. An obvious one is that the species in question are under extreme threat of genetic erosion or indeed of extinction. Adequate sampling and storage are indicated so that material can be provided for experimental studies and perhaps for later re-establishment under natural conditions. It might be felt that some species are not worth conserving but to the taxonomist, and especially the experimentalist, the conservation of a wide genetic base for evolutionary or other studies would always be an attractive proposition. This is the pure science argument.

There is then the economic argument, equally forceful, and perhaps even more so than the pure science one. It could be said that the species of certain families and genera are likely to be valuable in the future because other related ones have been so in the past, or that they themselves have shown promise as a result of preliminary assays. If the species itself has shown promise then it would rate highly for conservation. If it occurred in areas where others had shown promise, then, according to Vavilov's Law of Homologous Series (Vavilov, 1926), it should be preserved for that reason. The argument does not rest on the basis of assumed introgression or gene flow but because common selection pressures in any one region may have resulted in certain parallel adaptations being selected in unrelated species in the same area. The argument has not been applied to any extent in wild species but has had success with cultivated ones, since human or artificial selection seems to have been very intensive.

What decisions should we take regarding whole families? Would it be worth preserving all Leguminosae because of high protein content; or Gramineae because of the large number of valuable forage, pasture and cereal species in that family; or the families Solanaceae and Dioscoreaceae because of the steroid alkaloids they contain and which our drug industries cannot yet synthesize? My answer would be a cautious "yes", especially if at the same time one could add to these high priorities the others of actual or threatened genetic erosion.

Hegnauer (1975) has given considerable thought to this problem, stating that "It will never be possible to forecast the potential value of a taxon to different branches of science or to future plant utilization". He quotes a number of instances of the value of secondary metabolites in monotypic genera or specialized plants such as *Ginkgo*, *Sciadopitys* and *Lilaea*, which are rare and might well have become extinct before they could be studied.

So here again the taxonomist should assume an important role in drawing attention to materials that are rare or under threat, or which are related to others which possess actual or potential value to man. Beyond establishing these general guidelines it is difficult to go at present, but clearly those taxonomic studies indicating relationships will be of most value here.

The conservation of ornamentals

Finally, some mention should be made here of ornamentals. The same ideas, pressures and solutions apply to these as to the wild species discussed above (Creech, 1970). Many have been brought from the wild comparatively recently and hardly differ from the populations still existing in nature. A few show mutations such as variegated leaves, more petals, special colour forms and so on. Others are the result of crossing and selection between nearly related species whose history is well known. Others again—and these are very few— are perhaps more similar in their origin to our ancient crop plants. Roses, chrysanthemums, paeonies, tulips and a few others have a long period of cultivation whose beginnings lie in the history of the ancient civilizations of China, Persia, Egypt, Greece and Rome. Taxonomists have elucidated the origins of many of these ornamentals (see, for example, Wylie, 1952, 1954; Rowley, 1959), and can point to the wild prototypes, which are urgently in need of conservation. The conservation of the cultivars themselves will be touched on later.

RESEARCH ON THE CONSERVATION
OF GENETIC DIVERSITY

Various research topics have been mentioned previously. From an applied taxonomy view-point we still need to know more about population structure and sampling strategies, especially in relation to the sampling of vegetatively propagated crops (potatoes, yams, sweet potatoes, cassava, taro etc.) and to that of forest and fruit trees, where, according to Whitmore (1975b: 236–237), there is a density of generally not more than 13 to 24 individuals per 100 ha. This makes it necessary to preserve vast tracts of land for genetic reserves and to sample over similarly large areas if seeds are to be conserved with a reasonable hope of capturing 95% of the genetic diversity of a region.

Then again, we have touched on the problems of the conservation of recalcitrant seeds, which are produced by many forest and most fruit trees, as well as such essential economic crops as coffee, cocoa, rubber, oil palm, citrus fruits, tea and a large range of tropical fruits. Problems also exist in respect of crops that are largely vegetatively propagated and for which our knowledge of their reproductive biology is often very scanty. Research is under way on methods of long-term storage of meristems, but much remains to be done. Meristem storage of selected cultivars and of useful genetic stocks is also needed. Indeed it has been said that all genetic stocks of potatoes (of which some 12,000 are now being held in the living plant gene bank at the International Potato Center, Peru) should go into meristem culture. This seems excessive but it is a viewpoint that can be defended with some justice. If we consider the storage of genetic diversity of woody shrubs, even though research is still needed to solve certain problems, we could conveniently cut down a great deal on labour costs and avoid possible disease infection. Furthermore, the problems of international quarantine would be reduced considerably.

F

THE TAXONOMIST'S TASK

We have said enough in the preceding sections to show that more work by taxonomists is needed in connection with genetic resource exploration, evaluation and conservation. Even though the main outlines of the taxonomy of temperate crops are known this cannot be said for the tropical ones.

Especially worthy of study amongst the tropical crops are the vegetatively propagated ones, about which we know very little. Of course we can put a name to most of them but our understanding of their reproductive biology, their wild prototypes, or even their basic chromosome number, is very limited. We know little of the taxonomy of *Curcuma* (turmeric), *Zingiber* (ginger), *Xanthosoma* (Indian arrowroot), *Colocasia* (taro), *Amorphophallus* or *Dioscorea*. Even the wild prototypes of *Manihot* (cassava) and *Ipomoea* (sweet potato) are not known—a fact which Baron von Humboldt pointed out in 1807, just 170 years ago!

Since germ-plasm collections of our major crop plants are now being assembled on an impressive scale for genetic conservation purposes, a unique opportunity now exists for taxonomists to study them. The International Potato Center at Lima, Peru, possesses a collection of over 12,000 accessions of cultivated potatoes, classified into some seven distinct species; a counterpart collection of wild species amounting to several thousand accessions exists at Sturgeon Bay, Wisconsin. The USDA possesses an enormous collection of wheat cultivars and wild relatives at Beltsville, Maryland, whilst a comparably large collection of maize is held at the International Maize and Wheat Improvement Center near Mexico City. For rice, a very large collection of rice cultivars and population samples is held at the International Rice Research Institute in the Philippines; and sorghums, millets, chick peas, various species of *Phaseolus* beans, cassava, yams, sweet potatoes and many other important world food crops are being assembled and stored by international and national institutes.

This work is in many cases supported by international agencies such as FAO. For the last 3 years an International Board for Plant Genetic Resources has been in existence whose briefing is to support, by all means that lie in its power, the assemblage, storage, evaluation and use of the world's genetic resources of cultivated plants and their wild relatives. This body derives its funding from donor countries, via the World Bank and the Consultative Group on International Agricultural Research, and its interests in genetic resources are specifically directed towards helping Third World countries. Apart from this, nationally funded gene banks and research in progress on the material held in them are also playing a part in this enterprise. A few worthy of mention are the N.I. Vavilov Institute of Plant Industry at Leningrad, U.S.S.R., the gene bank at Braunschweig, West Germany, that at Gatersleben in East Germany, the Germplasm Laboratory in Bari, Italy and many others in Europe, Asia, Africa and the Americas.

So let me end this chapter by making a plea for more work by taxonomists on the crop plants, forest trees and fruit trees, as well as on related wild species, especially those of the tropics. The groundwork still needs to be done and the rewards are thereby much greater. All genetic resources exploration, conservation and evaluation work needs a sound foundation of taxonomy, which in many instances still needs to be laid. Let us hope that taxonomists will respond to this challenge.

REFERENCES

Baker, H. G. (1970). Taxonomy and the biological species concept in cultivated plants. In "Genetic Resources in Plants" (O. H. Frankel and E. Bennett, eds), IBP Handbook No. 11, pp. 49–68. Blackwell, Oxford and Edinburgh.

Bennett, E. (1970). Tactics of plant exploration. In "Genetic Resources in Plants" (O. H. Frankel and E. Bennett, eds), IBP Handbook No. 11, pp. 157–179. Blackwell, Oxford and Edinburgh.

Brenan, J. P. M., Ross, R. and Williams, J. T., eds (1975). "Computers in Botanical Collections", pp. 216. Plenum Press, London and New York.

Burley, J. and Styles, B. T., eds (1976). "Tropical Trees—Variation, Breeding and Conservation". Linnean Society Symposium Series No. 2. Academic Press, London and New York.

Creech, J. L. (1970). Tactics of exploration and collection. In "Genetic Resources in Plants" (O. H. Frankel and E. Bennett, eds), IBP Handbook No. 11, pp. 221–229. Blackwell, Oxford and Edinburgh.

D'Amato, F. (1975). The problem of genetic stability in plant tissue and cell cultures. In "Crop Genetic Resources for Today and Tomorrow" (O. H. Frankel and J. G. Hawkes, eds), IBP 2, pp. 333–348. Cambridge University Press, Cambridge.

Frankel, O. H. and Bennett, E., eds. (1970). "Genetic Resources in Plants—Their Exploration and Conservation", IBP Handbook No. 11. Blackwell Scientific Publications, Oxford and Edinburgh.

Frankel, O. H. and Hawkes, J. G., eds. (1975). "Crop Genetic Resources for Today and Tomorrow", IBP 2. Cambridge University Press, Cambridge.

Godron, M. and Poissonet, J. (1970). Standardization and treatment of ecological observations. *In* "Genetic Resources in Plants" (O. H. Frankel and E. Bennett, eds), IBP Handbook No. 11, pp. 189–204. Blackwell, Oxford and Edinburgh.

Gomez-Campo, C. (1972). Preservation of west Mediterranean members of the cruciferous tribe Brassiceae. *Biol. Conserv.* **4**, 355–360.

Gomez-Pompa, A., Toledo, J. A. and Soto, M. (1975). Data for the Flora of Veracruz program. *In* "Computers in Botanical Collections" (J. P. M. Brenan, R. Ross and J. T. Williams, eds), pp. 35–51. Plenum Press, London and New York.

de Guzman, E. D. (1975). Conservation of vanishing timber species in the Philippines. *In* "South-east Asian Plant Genetic Resources" (J. T. Williams, C. H. Lamoureux and N. Wulijarni-Soetjipto, eds), pp. 198–204. National Biological Institute, Bogor.

Harlan, J. R. (1966). Plant introduction and biosystematics. *In* "Plant Breeding" (K. J. Frey, ed.), pp. 55–83. Iowa State University Press, Ames.

Harlan, J. R. and de Wet, J. M. J. (1971). Towards a rational classification of cultivated plants. *Taxon* **20**, 509–517.

Harlan, J. R. and de Wet, J. M. J. (1972). A simple classification of cultivated sorghum. *Crop Sci.* **12**, 172–176.

Harrington, J. F. (1970). Seed and pollen storage for conservation of plant gene resources. *In* "Genetic Resources in Plants" (O. H. Frankel and E. Bennett, eds), IBP Handbook No. 11, pp. 501–521. Blackwell, Oxford and Edinburgh.

Hawkes, J. G. (1970). The taxonomy of cultivated plants. *In* "Genetic Resources in Plants" (O. H. Frankel and E. Bennett, eds), IBP Handbook No. 11, pp. 69–85. Blackwell, Oxford and Edinburgh.

Hawkes, J. G. (1976). Sampling gene pools. *In* "Conservation of Threatened Plants" (J. B. Simmons, R. I. Beyer, P. E. Brandham, G. Ll. Lucas and V. T. H. Parry, eds), pp. 145–154. Plenum Press, London and New York.

Hegnauer, R. (1975). Secondary metabolites and crop plants. *In* "Crop Genetic Resources for Today and Tomorrow" (O. H. Frankel and J. G. Hawkes, eds), IBP 2, pp. 249–265. Cambridge University Press, Cambridge.

Henshaw, G. G. (1975). Technical aspects of tissue culture storage for genetic conservation. *In* "Crop Genetic Resources for Today and Tomorrow" (O. H. Frankel and J. G. Hawkes, eds), IBP 2, pp. 349–357. Cambridge University Press, Cambridge.

Hersh, G. N. and Rogers, D. J. (1975). Documentation and information requirements for genetic resources application. In "Crop Genetic Resources for Today and Tomorrow" (O. H. Frankel and J. G. Hawkes, eds),IBP 2, pp. 407–446. Cambridge University Press, Cambridge.

Humboldt, Baron A. von (1807). "Essai sur la Geographie des Plantes". Paris.

Lamoureux, C. H. (1975). Tropical forests of South East Asia: genetic resources for plants other than timber. In "South-east Asian Plant Genetic Resources" (J. T. Williams, C. H. Lamoureux and N. Wulijarni-Soetjipto, eds), pp. 213–216. National Biological Institute, Bogor.

Lucas, G. Ll. (1976). Conservation: recent developments in international co-operation and legislation. In "Conservation of Threatened Plants" (J. B. Simmons, R. I. Beyer, P. E. Brandham, G. Ll. Lucas and V. T. H. Parry, eds), pp. 271–277. Plenum Press, London and New York.

Morel, G. (1975). Meristem culture techniques for the long-term storage of cultivated plants. In "Crop Genetic Resources for Today and Tomorrow" (O. H. Frankel and J. G. Hawkes, eds), IBP 2, pp. 327–332. Cambridge University Press, Cambridge.

Roberts, E. H. (1975). Problems of long-term storage of seed and pollen for genetic resources conservation. In "Crop Genetic Resources for Today and Tomorrow" (O. H. Frankel and J. G. Hawkes, eds), IBP 2, pp. 269–296. Cambridge University Press, Cambridge.

Rogers, D. J., Snoad, B. and Seidewitz, L. (1975). Documentation for genetic resources centers. In "Crop Genetic Resources for Today and Tomorrow" (O. H. Frankel and J. G. Hawkes, eds), IBP 2, pp. 399–405. Cambridge University Press, Cambridge.

Rowley, G. (1959). Ancestral China roses. J. R. hort. Soc. 84, 270–273.

Sakai, A. and Noshiro, M. (1975). Some factors contributing to the survival of crop seeds cooled to the temperature of liquid nitrogen. In "Crop Genetic Resources for Today and Tomorrow" (O. H. Frankel and J. G. Hawkes, eds), IBP 2, pp. 317–326. Cambridge University Press, Cambridge.

Sastrapradja, S. (1975). Tropical fruit germplasm in South East Asia. In "South-East Asian Plant Genetic Resources" (J. T. Williams, C. H. Lamoureux and N. Wulijarni-Soetjipto, eds), pp. 33–46. National Biological Institute, Bogor.

Simmons, J. B., Beyer, R. I., Brandham, P. E., Lucas, G. Ll. and Parry, V. T. H. (1976). "Conservation of Threatened Plants", Plenum Press, London and New York.

Sykes, J. T. (1975). Tree crops. In "Crop Genetic Resources for Today and Tomorrow" (O. H. Frankel and J. G. Hawkes, eds), IBP 2, pp. 111–115. Cambridge University Press, Cambridge.

Thompson, P. A. (1970). Seed banks as a means of improving the quality of seed lists. Taxon 19, 59–62.

Thompson, P. A. (1974). The use of seed banks for conservation of populations of species and ecotypes. *Biol. Conserv.* **6**, 15–19.

Thompson, P. A. (1975). The collection, maintenance and environmental importance of the genetic resouces of wild plants. *Environ. Conserv.* **2**, 223–228.

Thompson, P. A. (1976). Factors involved in the selection of plant resources for conservation as seed in gene banks. *Biol. Conserv.* **10**, 159–167.

Vavilov, N. I. (1926). (Studies on the origin of cultivated plants.) *Bull. appl. Bot. Genet. Pl. Breed.* **16**, 1–248.

Vavilov, N. I. (1940). The new systematics of cultivated plants. *In* "The New Systematics" (J. Huxley, ed.), pp. 549–566. Clarendon Press, Oxford.

Villiers, T. A. (1975). Genetic maintenance of seeds in imbibed storage. *In* "Crop Genetic Resources for Today and Tomorrow" (O. H. Frankel and J. G. Hawkes, eds), IBP 2, pp. 297–315. Cambridge University Press, Cambridge.

de Wet, J. M. J., Gray, J. R. and Harlan, J. R. (1976). Systematics of *Tripsacum* (Gramineae). *Phytologia* **33**, 203–227.

Whitmore, T. C. (1975a). South East Asian forests as an unexploited source of fast growing timber. *In* "South-east Asian Plant Genetic Resources" (J. T. Williams, C. H. Lamoureux and N. Wulijarni-Soetjipto, eds), pp. 205–212. National Biological Institute, Bogor.

Whitmore, T. C. (1975b). "Tropical Rain Forests of the Far East". Clarendon Press, Oxford.

Williams, J. T., Lamoureux, C. H. and Wulijarni-Soetjipto, N., eds. (1975). "South-east Asian Plant Genetic Resources". National Biological Institute, Bogor.

Wylie, A. P. (1952). The history of the garden narcissi. *Heredity* **6**, 137–156.

Wylie, A. P. (1954). The history of garden roses. Masters Memorial Lecture. *J. R. hort. Soc.* **79**, 555–571; **80**, 8–24, 77–87.

Zohary, D. (1970). Wild wheats. *In* "Genetic Resources in Plants" (O. H. Frankel and E. Bennett, eds), IBP Handbook No. 11, pp. 239–247. Blackwell, Oxford and Edinburgh.

Zohary, D. (1971). Origin of south-west Asiatic cereals; wheat, barley, oats and rye. *In* "Plant Life of South-west Asia" (P. H. Davis, P. C. Harper and I. C. Hedge, eds), pp. 235–263. Botanical Society of Edinburgh.

Zohary, D. (1972). The wild progenitor and the place of origin of the cultivated lentil: *Lens culinaris. Econ. Bot.* **26**, 326–332.

Note: A general bibliography of plant genetic resources has been published and, although not cited in the text, the reference is given herewith:

Hawkes, J. G., Williams, J. T. and Hanson, J. (1976). "A Bibliography of Plant Genetic Resources". FAO, Rome.

Chapter 8

Euphrasia: a taxonomically critical group with normal sexual reproduction

P. F. YEO

INTRODUCTION

Most of the major taxonomically critical groups in Europe have peculiarities in their breeding system. Usually apomixis is involved, sometimes accompanied by facultative sexuality in some stocks, permitting spasmodic hybridization and segregation. Polyploidy is nearly always present and modification of the normal sexual process is apparently a response (doubtless established by natural selection) to the sterility which polyploidy has brought about.

Euphrasia (family Scrophulariaceae, tribe Rhinantheae) was first monographed by Wettstein (1896), and the species-concept which he adopted has been adhered to by some authors up to the present day (Pugsley, 1930; Smejkal, 1963; Yeo, 1972). Although both Wettstein (1896) and Pugsley (1930) recognized interspecific hybrids in *Euphrasia*, suggestions that the genus might prove to be apomictic were in circulation until about 1950. This article presents facts and speculations about the European representatives of the genus.

Summary of the biosystematic situation

1. There are two chromosome levels, diploid and tetraploid ($2n = 22$ and 44) (Yeo, 1954).
2. On each level, but especially at the tetraploid level, there is a series of more or less ill-defined taxa, treated for nomenclatural convenience as species.

3. They are self-fertile (Yeo, 1966).

4. They range from mainly autogamous (inbreeding) to mainly allogamous (outbreeding) (Wettstein, 1896; Yeo, 1966).

5. They are interfertile; hybrids range from nearly sterile to highly fertile (Yeo, 1966).

6. They are ecogeographically more or less well characterized (Yeo, 1966, 1968).

7. They actually interbreed in nature where habitats mingle or where there is disturbance (as reported by many observers, cf. Yeo, 1975).

8. Because of the weakness of the taxonomic distinctions between the most similar species and the aggravation of this situation by hybridization, they are semispecies in the sense of Grant (1971, his p. 48).

9. The morphological range has wide limits (Wettstein, 1896), so it is difficult to keep the taxa to a very small number, but the alternative seems to be to make the taxa very "small" and numerous; intermediate levels of taxonomic splitting (e.g. Hartl, 1972) seem unnatural.

Special features of the group

1. They are green facultative hemiparasites, in common with the rest of the Rhinantheae; root hairs are few and the slender roots form haustorial connections with the roots of host-plants [the transfer of nutrients from host to parasite has been studied in the closely related genus *Odontites* (Govier et al., 1967, 1968)].

2. Separate patterns of variation in habit characters of an ecologically adaptive nature (Fig. 1) and in taxonomic characters similar to those of other groups, where taxonomic differentiation is clearer, are superimposed and are often further complicated by local and clinal variation.

3. Most of the species occupy more or less closed grassland habitats, which is unusual for annual or biennial plants (Karlsson, 1974).

4. There is evidence for hybridization across the "ploidy" barrier, at least from tetraploid to diploid; the results are either introgression or speciation (Yeo, 1956).

5. Species-distinctness varies on a regional basis; there are some very distinct species around the north end of the Adriatic Sea; in the

Alpine region species-distinctions are also quite good; in north-west Europe there are more species than in, say, the western part of the Alps, but they are less distinct; in Iceland there is an evident pattern of variation but it is not possible to provide a workable taxonomy comparable to that applicable in the British Isles and Fennoscandia.

The problems posed

1. Is there any host-specificity which could affect the taxonomic situation?
2. Can the results of the superimposed patterns of variation be encompassed by formal taxonomy?
3. How far can we account for even the rather poor level of species-differentiation which exists in glaciated areas of North Europe, and why are the principal species of the Alps more distinct than those of North Europe?
4. Why does the breeding system vary?
5. What suggestions can be made about the evolutionary development of *Euphrasia* as a polyploid complex in Europe (a polyploid complex being defined as comprising more than one level of "ploidy")?

CONSIDERATION OF THE PROBLEMS

Host-specificity

The observations of Yeo (1964) and earlier workers reviewed by him suggested that most *Euphrasia* species could benefit parasitically from attachment to a wide range of host species, including some with which they would never come into contact naturally; Yeo encountered one "host" species, however, which failed to benefit every one of the wide variety of *Euphrasia* samples tested. Wide variation was found in the vigour of *Euphrasia* plants on different hosts, and there were differences in the responses of different *Euphrasia* species, and even of different samples of the same species. However, since it was evident that there was no close specialization of particular *Euphrasia* species on particular hosts, it was clear that this in itself could not account for the large number of semi-species in the genus.

A question which Yeo's experiments were not designed to answer was whether individuals of a *Euphrasia* population might differ in parasitic ability. The difficulty here is that *Euphrasia* cannot, as far as is known, be propagated vegetatively. However, a positive answer to this question is implicit in the results of Atsatt and Strong (1970) and Atsatt (1970b), working in California with the annual genus *Orthocarpus*, also of the Rhinantheae. The obligatorily outbreeding *O. purpurascens* Benth. was found to be capable, as is *Euphrasia*, of maturing autotrophically. When tested with six different host species it was less fecund with one of them than when growing auto-trophically. With the other five species there was enhancement of mean fecundity (measured by inflorescence length) ranging from slight to substantial. The range of variation also increased markedly with the mean, so that some individuals were remarkably vigorous, while others fell within the range of autotrophic plants. This clearly demonstrated genetic variation in the parasite in its response to various host species.

Atsatt and Strong (1970) grew a sample of *Orthocarpus* autotrophically and crossed the two strongest and the two weakest individuals. The progeny of the crosses were grown autotrophically; those from the strong individuals showed improved average performance, while those from the weak ones produced only two flowering plants from 200 seeds. Both F_2s approached the mean of the parental sample. Samples from the parental and F_2 generations were grown also in each case with hosts of two favourable species; the F_2 from the "strong" autotroph cross was more vigorous than the parental, and that of the "weak" very much more so. The two survivors in the "weak" line therefore carried genes which could recombine to produce markedly superior heterotrophs. The performance of the F_2s showed that autotrophic ability is not maximized by the selective forces operating in the field.

For plants of *Orthocarpus purpurascens* with physiological deficiencies, such as achlorophylly (Atsatt, 1970a), there would be a large element of chance in their success or failure to find hosts which could compensate for their deficiencies and so allow them to repro-duce. Atsatt (1970a) described this compensation as "chemical buffering" or "canalization", but it differs from that described by

Waddington (1975) in that the substances which allow normalization of development in the presence of an aberrant genotype come from the host and depend on the host's genotype.

Atsatt's forthcoming publications will show that haustoria in *Orthocarpus* form in response to chemical stimuli, and that heterotrophic ability is directly related to haustoria-forming ability, and also to the host's capacity to induce the formation of haustoria (P. R. Atsatt, personal communication).

Orthocarpus purpurascens grows in California in "annual grasslands", which show high species-diversity and low dominance-diversity in any one year, but marked changes of dominants from year to year. The variation in *O. purpurascens* in respect of autotrophic ability and individual pre-adaptation to different host-species would seem to suit it well to the peculiarities of the host-environment. Like *Euphrasia*, however, the self-incompatible species of *Orthocarpus* show considerable and often confusing infraspecific variation; they hybridize quite frequently and commonly show introgressive effects (Atsatt, 1970b). It is suggested that local patterns of variability may depend on the complex chemical environment provided by a wide range of grassland hosts, which permits physiological buffering and allows the expression and reproduction of a wide range of genotypes. In outcrossing populations such relaxation of selection should generally increase levels of genetic variability, although canalization may diminish its phenotypic expression (P. R. Atsatt, personal communication).

The inbreeding species of *Orthocarpus* were also investigated by Atsatt (1970b). These were only found where outbreeders were present, and they often occurred with sparse uniform distribution, in contrast to a marked clustering of the outbreeders. It is suspected that they may not possess the diversity of host-adaptation that the outbreeders have, and that good dispersal may enable them to find suitable hosts and at the same time result in their relatively uniform population densities. However, it is also suspected (p. 610 in Atsatt, 1970b) that occasional crossing with outbreeders may have augmented the pool of genetic variability in inbreeders, though it may be partitioned into relatively homogeneous family groups. It was thought that the striking variability of self-incompatible *Orthocarpus* species, in the response of individuals to potentially beneficial hosts, was related to their outbreeding habit. The variation in flower size and degree of dichogamy (asynchrony of ripening of male and female organs) in *Euphrasia* reported by Wettstein (1896) and Yeo (1966) were seen as an indication that control of recombination is important here too, and might affect the genetics of individual reactions to host-species.

In fact, the results of a trial of *Euphrasia nemorosa* (Pers.) Wallr. described by Yeo (1964: 11–13) are similar to those for *Orthocarpus purpurascens*. The criterion of growth was dry weight, not number of flowering nodes. When the results are presented as histograms (as in Yeo, 1959, p. 124) they look very similar to those of Atsatt and Strong (1970) (except for the omission of the class of complete failures, i.e. those individuals which died between potting-up on 8–10 April and the start of regular recording on 8 May). Thus the two hosts which gave greatly enhanced growth also produced very high variability. If the *E. nemorosa* trial represents an expression of genetic variation in the population, then the genetical explanation of its occurrence may differ from that offered for *Orthocarpus*. *Euphrasia* is self-compatible, and *E. nemorosa* has small to medium-sized corollas (within the *Euphrasia* range). These flowers would readily self-pollinate and would not, on the basis of my general experience (p. 241 in Yeo, 1966), be expected to receive many insect visits. *Euphrasia* tends to grow in permanent grassland with reasonably stable composition from year to year. Therefore, what may be needed is simply a reasonably diverse but stable mix of genotypes, and this may be attained by a breeding system different from that of *Orthocarpus*, where a constant reshuffling of genes (which would certainly carry a penalty in wasted genotypes) imposed by self-incompatibility, is the most effective method of coping with the fluctuating host-environment. In the *Festuca (Vulpia) microstachys* Nutt. complex studied by Kannenberg and Allard (1967), cleistogamy has led to individual homozygosity, but a large element of genetic variation persists in populations. This group provides, therefore, an example of the maintenance of a high level of genetic diversity without a high level of outbreeding. In fact, the habitats of the different *Euphrasia* species vary in the extent to which they are closed or open, and in host-species composition from site to site. The situation shown by Atsatt's work is an extreme one but it is a pointer to possible studies on the variation of heterotrophic capabilities within *Euphrasia* populations and its links with the pronounced variation in the breeding system, as suggested by Atsatt (1970b: 610).

Kinds of variation to be given priority in taxonomy

Firstly, I consider there is no fundamental reason why obviously ecologically adaptive characters should not be used in taxonomy. If they relate to some constant feature of the organism's mode of life they are particularly valuable in taxonomy because they have to be present if the organism is to function properly. Differences between related organisms in such characters are indicative of differences in mode of life and are therefore particularly significant indicators of interspecific divergence. On the other hand, those characters which are plastic in response to variation in the environment are of little use.

In practice we find in *Euphrasia* and allied genera much variation that is ecologically adaptive. The illustration (Fig. 1) from Karlsson (1974) is an example of this; in this case the variation shown is within the species as understood in *Euphrasia,* and the variation is parallel within the different species. However, it will be seen that, of the five habitats, only one is inhabited by all four species, and that each species has some ecological limitations. If the characters visible in the illustration are ecotypic in nature other characters must be available for distinguishing the species. Before considering these, however, it must be admitted that while ecologically adaptive characters are in some instances regarded as infraspecific, in others they are used to characterize species. For example, there is a series of characters associated with habitats where the growing season is short, and these may be found in a number of genera of the Rhinantheae (excellently surveyed by Karlsson, 1974). These characters, and their adaptive significance, are very clearly brought out for *Rhinanthus serotinus* (Schönheit) Oborny in the Netherlands by Borg (1972); they are broad leaves, long internodes, low node of flowering, large capsules and large and numerous seeds. Short growing seasons for Rhinantheae are caused in the Netherlands, and many other areas, by mowing for hay. In unmown but adjacent habitats, converse characters are found. Such contrasts gave rise to Wettstein's (1901) concept of seasonal dimorphism. According to this, flowering and fruiting can take place before the mowing or after. The two forms ought to be able to occur in the same taxon at the same site. Wettstein demonstrated this, but in fact it seems to be a rare phenomenon. Habitats with growing seasons of different duration also occur in close proximity in dissected mountain areas such as the Alps, and there, similar ecotypic or, in fact, ecoclinal

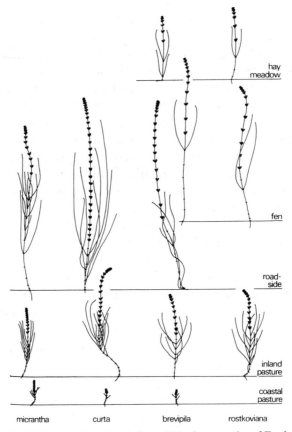

FIG. 1. Parallel, habitat-correlated variation within four species of *Euphrasia*. Average specimens collected from the habitats in southern Sweden, province of Skåne. Nodes, branches and flowers are represented. From Karlsson (1976) with kind permission of the author and the publishers of *Botaniska Notiser*.

variation is then frequent. The use of such "integrated adaptational character complexes" in separating species is regarded as inadmissible by Karlsson (1976); he considers that they characterize mere ecotypes. Yeo (1972) did not give infraspecific recognition to this kind of variation in *Euphrasia stricta* D. Wolff ex J. F. Lehm. as it does not appear very extreme, but in *E. rostkoviana* Hayne he accepted three rather arbitrarily defined subspecies.

The use of these short-season adaptations as characters of a *species* is best exemplified by *Euphrasia frigida* Pugsley, which is a North European mountain plant. Some of these characters are also seen in the corresponding species of the Alps, namely *E. minima* Jacq. ex

DC. These are distinct montane taxa, and the characters of broad foliage, low node of flowering and large capsules can here be used for taxonomic purposes. Similarly, habit characters which appear to be adaptive to conditions of close grazing sometimes appear constant in a species, as in *E. confusa* Pugsley, but they also recur in other species, such as *E. nemorosa* and *E. stricta*, in grazed situations.

So here, as in the set of species illustrated by Karlsson, we must have other characters which we can use to distinguish the species. In general such additional characters are those of leaf-outline and leaf-toothing, corolla size and flower colour and, in some cases, a range in capsule shapes differing from that of other taxa. Finally there is the local and regional differentiation that can sometimes be seen. The two species in which it is perhaps best developed are *E. rostkoviana*, in which it has led to the proposal of *E. fennica* Kihlman as a distinct species, and *E. salisburgensis* Funck, from which *E. lapponica* T. C. E. Fries was segregated. In the areas from which *E. fennica* and *E. lapponica* have been recognized, more typical forms of the supposed ancestral species also occur, and the taxa themselves vary ecotypically in respect of flowering date and the associated characters already mentioned.

In Yeo's taxonomic treatments (Yeo, 1972, and unpublished), the variation of the second type, that is the group including leaf shape, flower size and flower colour, has been used as a primary basis, and where the more obviously adaptive habit characters show good correlation with this they have been used too. The regional and local variation, though it involves characters of a type more "taxonomic" than ecotypic, has been used very little. This is because of the frequent geographical overlap of the more generally distributed variants and the lack of uniformity caused by seasonal-ecotypic variation. However, some local variants at varietal rank will be recognized in forthcoming taxonomic publications.

In conclusion, I would say that my practice in *Euphrasia* taxonomy has been simply to look for characters that "work", which is what taxonomists have traditionally done. However, the way in which I have selected characters from the three types mentioned might be challenged on grounds of consistency or logicality, and alternative priorities might conceivably be adopted.

Karlsson (1976) proposes applying the additional criterion of a substantial degree of genetic isolation for recognition at the rank of species. In this way he justifies recognition of only five entities as species in Sweden. One of these, the *E. stricta* complex, includes a wide range of morphological types which I have treated as species; but, as stated in the introductory summary of the biosystematic situation, such intermediate levels of lumping give unnatural-looking results.

Level of species-differentiation and its maintenance

Some years ago Yeo (1968) considered possible explanations for the multiplicity of *Euphrasia* species, and their distinctness, such as it is, in Europe. In view of the amount of glacial disturbance which the British Isles had suffered, it seemed difficult to account for the retention of even the existing degree of distinctness of so many taxa there. Some are not endemic and must have achieved their present distribution by post-glacial immigration. Some of the endemics are also relatively wide-ranging and must have undergone considerable migration in the post-glacial period. In view of this, the weakness of sterility barriers in the group is difficult to reconcile with the recurrence of particular taxa over wide areas of the British Isles. Two factors were put forward as accounting for the situation. Firstly, it was suggested that the favoured habitat-types of the various *Euphrasia* species must have extended back in time to the early post-glacial, and that the Euphrasiae spread as parts of migrating plant communities ("habitat stability"). Secondly, hybrids of the F_2 and later generations may come to comprise increasing proportions of individuals resembling one or other parent ("genetic coherence"); Anderson (1949) illustrated this graphically by his "recombination spindle". To account for the numerical abundance of taxa, Yeo (1968) suggested that, for reasons unspecified, colonization of different habitats might require closer specialization than in other groups.

Another possible explanation of the situation is that out of a mêlée of forms created by the action of the last glaciation on survivors from previous glaciations and interglacials, ecotypes were selected out in the post-glacial as the vegetation settled down. However, Yeo (1968) considers that *Euphrasia* species rank higher than the ecotypes of Turesson, and are more equivalent to those of Clausen and his school. Some of the *Potentilla* ecotypes of Clausen *et al.* (1940) have been recognized as species, subspecies or varieties in western North American Floras published subsequently (Abrams, 1944; Munz, 1959; Hitchcock and Cronquist, 1961).

For the close specialization to habitat which seems to be required in *Euphrasia*, an explanation is offered by Karlsson (1974: 531). In considering recurrent (i.e. parallel) ecotypic differentiation in different taxa of the Rhinantheae and of *Gentianella*, he suggests that it is because the plants are hapaxanthic (annual or biennial) in habitats composed largely of perennials that "a series of conspicuous ecotypes" arises. Their dependence on seed-setting demands good adaptation, and their relatively short generation time (bearing in mind low rates of seedling establishment of perennials in closed communities) makes this possible. Both groups of hapaxanths are enabled to exist in closed communities by their irregular nutrition: parasitism in Rhinantheae, and mycorrhizal formation in *Gentianella*.

Both Karlsson (1974: 537) and Yeo (1968) have emphasized the importance of post-glacial habitat disturbance and the opportunities which it provided for hybridization and segregation. What has impressed me is that despite this, a workable, if difficult, *Euphrasia* taxonomy can be attained. Karlsson is more inclined than I am to attribute present-day variation patterns to ecotypism in the sense of Turesson (which produces parallel forms polytopically). He also emphasizes the extent to which grassland habitats are the result of human activity, from which it follows that some of the ecotypes are of quite recent origin. In fact, the British Isles probably have a wider range of natural grassland habitats than Sweden and Denmark, if not Norway, on account of their varied geology, dissected coasts, exposure to Atlantic winds and wet hill country. Karlsson and Yeo are therefore probably looking at situations which differ in their balance between mere ecotypism and something approaching normal speciation.

Even if it is granted that special explanations are required to account for a pattern of variation in northern Europe that permits recognition of numerous species, it remains a fact that the species of the Alps (the only other area of significant species-richness of *Euphrasia* in Europe) are more distinct than those of northern Europe. It seems probable that before the glaciations a few tetraploids existed in the Arctic, while in the Alps there were rather numerous and diverse diploids and tetraploids. The greater level of distinctness among the Alpine species suggests that they may be older.

The effects of the Pleistocene glaciations would be as follows:

1. Great ecological disturbance in the Alps, showing itself mainly as altitudinal oscillations of climatic zones,

2. Origin of new tetraploids made possible by this disturbance,

3. A meeting of emigrant Arctic species with these native Alpine species, with possibilities of new hybridization on the tetraploid level,

4. Ecological disturbance of the lower lands north of the Alps resulting not only in altitudinal oscillation but in extensive lateral migrations; these would provide further opportunities for hybridization and differentiation, and opportunities for the northward spread of new tetraploids in times of climatic amelioration,

5. The restriction, by partial extinction, of many of the more distinct and ancient *Euphrasia* species to the well known refugial area in the mountains around the north end of the Adriatic (in the case of *Euphrasia*, the south-east Alps and the east central Alps, as defined by Pawlowski (1970, his Fig. 5), of which the first is the richer in endemics).

The inferior level of species-distinctness in northern Europe could, then, be due to paucity of surviving ancient species, and the effects of climatic change listed above. On the other hand, in Iceland a workable taxonomy is not really possible for the commonest forms. If my interpretation of the specimens is correct, two very small-flowered species, *E. scottica* Wettst. and *E. ostenfeldii* (Pugsley) Yeo, are accompanied by unusually small-flowered states of *E. frigida*. But only a small proportion of the material can be assigned to these species. There seems to be an abundance of apparently hybrid populations, usually morphologically uniform. Small flower sizes are an indication of prevalent inbreeding, which must have led to uniformity subsequent to occasional hybridization. Thus we have in Iceland the paradox that taxonomic breakdown is caused by hybridization in plants which are the least well adapted to outbreeding. The same phenomenon probably accounts for the very similar type of variation in the small-flowered *E. minima* of the Alps (Fig. 2). The similarity of this outcome to the production of "Jordanons" (more or less pure lines sometimes recognized as species) in weedy annuals is due to the inbreeding component which is common to both situations.

Variation in amount of outbreeding

It was shown by Wettstein (1896), in observations later supported by Yeo (1966), that small-flowered Euphrasiae are readily, and even primarily, self-pollinated. The larger the corolla, however, the greater the degree of protogyny (female organs ripening earlier than male), and the more remote is the stigma from the stamens, so that the chances of cross-pollination are progressively greater until, in the largest-flowered species, self-pollination is rather difficult.

Among European species, corolla length ranges from about 3 mm in *E. bottnica* Kihlman to about 12 mm in *E. alpina* Lam. and some populations of *E. rostkoviana*. Each species has its own characteristic range within this wide span, though a few common and widespread species are exceptionally variable. This situation suggests that the control of recombination is important to *Euphrasia*, as pointed out by Atsatt (1970b: 610) who, in *Orthocarpus*, found the more extreme differences described earlier. Two possible reasons for having small corollas were suggested by Yeo (1968). The more general one was that a newly evolved species might be enabled to spread through an area occupied by another species if it had a degree of genetical isolation provided by a small-sized corolla. The other reason was a specialized one applicable only to *E. micrantha* Reichenb. Flowering shoots of this plant show a remarkable resemblance to those of *Calluna vulgaris* (L.) Hull with which *E. micrantha* is almost invariably associated. Possibly this is mimetic in relation to pollination, in which case the small size is incidental, and part of a character-complex actually related to outbreeding. But even here, pollen is shed before the flower opens (Yeo, 1966). A third, and rather obvious cause for small corolla size, one classically recognized, is a habitat poor in insect life. Kevan (1972) has shown that pollinating insects may be quite abundant in the Arctic but that local shortages, related to habitat conditions, occur. This could account for small flowers in the Icelandic taxa *E. frigida*, *E. ostenfeldii* and *E. scottica*, which also occur in wet and windy areas of the northern British Isles or Norway, and in *E. minima* of the Alps. At the same time it can account for the remarkably large corollas of some populations of *E. saamica* Juz. from Arctic Europe, and the large-flowered *E. alpina*, which has its main altitudinal range between 1600 m and 2200 m. Even so, since the range for *E. minima*, 1500–2500 m, is comparable to that of *E. alpina* (Hartl, 1972), it seems that several factors may be involved, and that the control of variation of individual parasitic adaptation, stressed by Atsatt (1970b: 610), could well be one of them. Phytosociological studies of *Euphrasia* habitats combined with progeny-testing would be needed to investigate this possibility.

I have the impression from a summertime visit to the Alps in 1968 that two large-flowered species, *E. rostkoviana*, inhabiting closed grassland habitats, and *E. alpina*, from generally less closed habitats, are more variable than smaller-flowered species, particularly within populations, and this is very likely related to the outbreeding habit (Fig. 2).

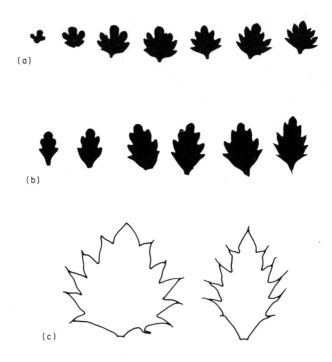

FIG. 2. Variation in leaf-shape within and between populations of *Euphrasia*. (a) Leaf spectrum of a plant of *E. minima*, Pont de Nant, Bex, Vaud, Switzerland, from a colony not near to any *E. salisburgensis*. (b) The same, from a separate colony closely associated with *E. salisburgensis* (see Fig. 3 for a leaf of that species). (c) Leaves from two cultivated plants of *E. alpina*; first generation from wild-collected seed (there is a difference of 1 in the node number from which the leaves were taken, but this could account for only a little of the difference seen). From Simplon Pass, Valais, Switzerland, *c.* 1380 m alt.

E. hirtella Jordan ex Reuter and *E. rostkoviana*, which are apparently closely related, show an extreme divergence in corolla size. I saw them growing near each other at several sites in Vaud and Valais in 1968; perhaps here the corolla difference is primarily an isolating mechanism, but it cannot be without effect on the breeding system.

Possible interrelationships of diploids and tetraploids

I am strongly inclined to support, in the case of *Euphrasia*, Atsatt's view (p. 611 in Atsatt, 1970b) that the retention of capacity for inter-specific gene exchange is important in the evolution of the group and in promoting its adaptive variation. In *Euphrasia* there is probably a rough correspondence between cross-compatibility and morpho-logical difference, which itself probably corresponds at least approx-imately with the lapse of time since the divergence of the species. I also think that crosses which are difficult for reasons of poor cross-compatibility, or are highly sterile in their progeny, may have been of great importance in the evolution of new lines in *Euphrasia* (Yeo, 1956, 1975), even without a change of chromosome number. It is only the 100% sterility barrier which can prevent genetic interaction between different taxa (p. 51 in Grant, 1971).

The evidence available for consideration of allopolyploidy in European *Euphrasia* is as follows:

1. We know quite a lot of chromosome numbers, and there are diploids ($2n = 22$) and tetraploids ($2n = 44$) in both the major groups, namely subsections Angustifoliae and Ciliatae (von Witsch, 1932; Yeo, 1954, 1956, 1970; Favarger, 1969; Feoli and Cusma, 1974).
2. A naturally occurring triploid individual, *E. anglica* Pugsley × *E. micrantha*, found by Yeo in England, had at meiosis approximately 11 bivalents and 11 univalents (Yeo, 1956), but also occasional trivalents.
3. An artificial hybrid between members of the two subsections, namely *E. tetraquetra* (Bréb.) Arrondeau × *E. salisburgensis*, produced meiotic pairing figures centred on 10 and 11 bivalents (Yeo, 1966), with consequent high sterility.
4. Crosses among tetraploids within subsection Ciliatae are usually rather fertile, but if rather dissimilar parents are used, some meiotic pairing failure occurs (Yeo, 1966).
5. *E. alpina* was artificially crossed with *E. rostkoviana* and *E. picta* Wimmer, which are closely related to one another; meiosis in *E. alpina* × *E. rostkoviana* F_1 showed 8–11 bivalents, and in *E. alpina* × *E. picta* 1–10 bivalents (all parents diploid, $2n = 22$) (Yeo, 1976).

A hypothetical scheme for the origin of certain groups and species has been drawn up which is consistent with the existing evidence (Fig.

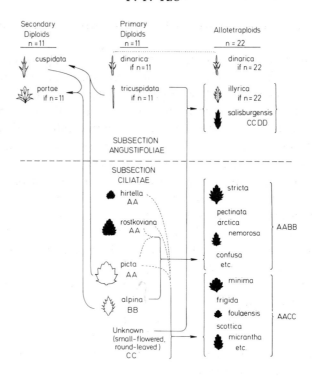

FIG. 3. Hypothetical scheme showing some possible genomic and descent relationships in *Euphrasia*. (No significance attaches to the varying ways of representing leaves; each leaf represents the species named to the right of it.)

3). Half-way down the central column are three species shown as having the *AA* genome; they belong to the Series Grandiflorae. It has been necessary to assume that the genomes are sufficiently similar for secondary diploids sometimes to be formed, but sufficiently distinct that on occasions when chromosome doubling follows a cross, the progeny can sometimes function as allotetraploids. Owing to the partial similarity of the parent genomes, these come into the category of segmental allopolyploids. However, genome *D* must be regarded as more distinct, to account for the high sterility of crosses between *E. salisburgensis* and members of subsection Ciliatae. If this is so it presents an obstacle to the proposed parentage for *E. cuspidata* Host (*top left*), which is admittedly based on a hunch about the morphology of the plant. A degree of similarity between the *B* and *C* genomes will facilitate crossing between the two main groups of tetraploids in subsection Ciliatae. Whether all the examples included are really primary tetraploids is not known. Many tetraploids omitted from the diagram are probably secondary, having arisen by hybridization among the primary tetraploid Ciliatae.

A further complication is gene exchange between the different chromosome levels. Yeo (1956) suggested that this has been responsible for introgressive modification of the diploid, *E. anglica,* in Britain; for the origin from *E. anglica* of a new species, *E. vigursii* Davey; and for the origin of *E. christii* Favrat from *E. alpina* (Yeo, 1970). A great obstacle to genomic studies so far has been the lack of any artificial triploid hybrids.

The picture in Fig. 3 differs from the hypothesis of Yeo (1956) in showing two separate groups of tetraploid Ciliatae with differing origins, whereas formerly a single or multiple origin from only one diploid pair was suggested. The present scheme is better from a morphological point of view but it postulates a further unknown diploid group.

An important feature of the scheme is that it shows interaction between the two European subsections of the genus leading to diversification within the Angustifoliae, and the generation of one widespread and successful species, *E. salisburgensis.* The rest of the Angustifoliae are confined to the area around the northern end of the Adriatic, and they comprise all the morphologically highly distinctive species in this area that have already been referred to in passing. It will be interesting to learn whether or not *E. salisburgensis* is the only tetraploid in the subsection.

The origin of new species by hybridization without change of chromosome number is a type of event postulated repeatedly over a long period of time; Grant gives examples (pp. 194–196 in Grant, 1971). Should such new hybrid species exhibit strong sterility barriers with the parent species the process is described as recombinational speciation (see Chapter 14 in Grant, 1971). Insufficient is yet known about *Euphrasia* to know for certain whether we have any species of homoploid (same chromosome number) hybrid origin, let alone what their genetic interactions might be.

Grant (1971) devotes a chapter to each of his concepts, "the polyploid complex" and "the homogamic complex". Obviously *Euphrasia* in Europe is a polyploid complex, and it seems that the diploids, as in the example of *Phacelia* investigated by Heckard (p. 306 in Grant, 1971), are more distinct (if we exclude a few minor endemics) than the tetraploids, and they may well turn out to have higher sterility barriers than the very low ones already known among some tetraploids (Yeo, 1966).

An example cited by Grant has a near parallel in Yeo's (1976) speculations on *Euphrasia alpina* and *E. rostkoviana*. Yeo pointed out a resemblance between the artificial F_1 *E. alpina* × *E. rostkoviana* and forms of *E. stricta*, such that the last might be an allopolyploid derived from the first two. At the same time natural *E. alpina* populations modified in the direction of *E. rostkoviana* were mentioned. Grant (1971: 307) describes how in *Achillea* Ehrendorfer found that two species, *A. asplenifolia* Vent. and *A. setacea* Waldst. & Kit. have formed an amphiploid (*A. collina* J. Becker ex Reichenb.), but that there has also been introgression from *A. setacea* into *A. asplenifolia*.

On the whole the biosystematic situation of *Euphrasia* in Europe is rather unlike that of other genera. It is true that the pattern of ecotypic differentiation is closely paralleled in other European Rhinantheae, but there are in addition sets of comparatively distinct and very indistinct species which are scarcely paralleled even in *Rhinanthus* and *Odontites* (including *Orthantha*). The work of Atsatt and Strong has contributed to our understanding of the genetic systems that may be promoted by the hemiparasitic habit, and Karlsson's and Borg's work have helped us to understand the processes of ecotypic differentiation and the significance of various characters of the ecotypes. But apart from this it seems to be easier to find examples of sexually reproducing groups of comparable taxonomic complexity in western North America.

For the future increase in our knowledge of *Euphrasia* we can look first to the publication of Karlsson's detailed studies of Swedish material, in which the population studies will be much more extensive than those of Yeo (1962), and cytotaxonomic studies that have at last been begun in the eastern Alps. The crosses made by Yeo (1966) in the attempt to produce triploids were biased in the direction (tetraploid on to diploid) less likely to be successful, so there is every reason for future workers to pursue this, and to follow up possibilities for the hybrid origin of species such as those suggested in Fig. 3. It is also clear that scope exists for a mor subtle approach to genotypic variation in parasitic abilities in th genus.

REFERENCES

Abrams, L. (1944). "Illustrated Flora of the Pacific States", vol. 2. Stanford University Press, Stanford.

Anderson, E. (1949). "Introgressive Hybridization". Wiley, New York and London.

Atsatt, P. R. (1970a). Hemiparasitic flowering plants: phenotypic canalization by hosts. *Nature, Lond.* **225**, 1161–1163.

Atsatt, P. R. (1970b). The population biology of annual grassland hemiparasites. II. Reproductive patterns in *Orthocarpus. Evolution* **24**, 598–612.

Atsatt, P. R. and Strong, D. R. (1970). The population biology of annual grassland hemiparasites. I. The host environment. *Evolution* **24**, 278–291.

Borg, S. J. ter (1972). "Variability of *Rhinanthus serotinus* (Schönh.) Oborny in Relation to the Environment". Rijksuniversiteit, Groningen.

Clausen, J., Keck, D. D. and Hiesey, W. M. (1940). Experimental studies on the nature of species. 1. The effect of varied environments on western North American plants. *Publs. Carnegie Instn* **520**.

Favarger, C. (1969). Notes de caryologie alpine, 5. *Bull. Soc. neuchâtel. Sci. nat.*, **92**, 13–30.

Feoli, E. and Cusma, T. (1974). Sulla posizione sistematica di *Euphrasia marchesettii* Wettst. *Giornale botanico italiano* **108**, 145–154.

Govier, R. N., Nelson, M. D. and Pate, J. S. (1967). Hemiparasitic nutrition in Angiosperms. I. The Transfer of organic compounds from host to *Odontites verna* (Bell.) Dum. (Scrophulariaceae). *New Phytol.* **66**, 285–297.

Govier, R. N., Nelson, M. D. and Pate, J. S. (1968). Hemiparasitic nutrition in Angiosperms. II. Root haustoria and leaf glands of *Odontites verna* (Bell.) Dum. and their relevance to the abstraction of solutes from the host. *New Phytol.* **67**, 963–972.

Grant, V. (1971). "Plant Speciation". Columbia University Press, New York.

Hartl, D. (1972). *Euphrasia. In* "Illustrierte Flora von Mitteleuropa" (G. Hegi, ed.), 2nd edition, Vol. 6, pp. 335–373. Carl Hanser, München.

Hitchcock, C. L. and Cronquist, A. (1961). "Vascular Plants of the Pacific Northwest", Part 3. University of Washington Press, Seattle.

Kannenberg, L. W. and Allard, R. W. (1967). Population studies in predominantly self-pollinated species. VIII. Genetic variability in the *Festuca microstachys* complex. *Evolution* **21**, 227–240.

Karlsson, Th. (1974). Recurrent ecotypic variation in Rhinantheae and Gentianaceae in relation to hemiparasitism and mycotrophy. *Bot. Notiser* **127**, 527–539.

Karlsson, Th. (1976). *Euphrasia* in Sweden: hybridization, parallelism, and species concept. *Bot. Notiser* **129**, 49–60.

Kevan, P. G. (1972). Insect pollination of high arctic flowers. *J. Ecol.* **60**, 831–847.

Munz, P. A. (1959). "A California Flora". University of California Press, Berkeley and Los Angeles.

Pawlowski, B. (1970). Remarques sur l'endémisme dans la flore des Alpes et des Carpates. *Vegetatio* 21, 181–243.

Pugsley, H. W. (1930). A revision of the British Euphrasiae. *J. Linn. Soc. Lond. (Bot.)* 48, 467–544.

Smejkal, M. (1963). Taxonomická studie československých druhů rodu *Euphrasia* L. *Biologické Práce* IX (9), 5–83.

Waddington, C. H. (1975). "The Evolution of an Evolutionist", Chapter 3. Edinburgh University Press, Edinburgh.

Wettstein, R. von (1896). "Monographie der Gattung *Euphrasia.*" Wilhelm Engelmann, Leipzig.

Wettstein, R. von (1901). Descendenztheoretische Untersuchungen über den Saisondimorphismus im Pflanzenreiche. *Denkschr. Akad. Wiss. Math.-Nat. Kl. (Wien)* 70, 305–346.

Witsch, H. von (1932). Chromosomenstudien an mitteleuropäischen Rhinantheen. *Oesterr. Bot. Zeitschr.* 81, 108–141.

Yeo, P. F. (1954). The cytology of British species of *Euphrasia. Watsonia* 3, 101–108.

Yeo, P. F. (1956). Hybridization between diploid and tetraploid species of *Euphrasia. Watsonia* 3, 253–269.

Yeo, P. F. (1959) Experimental taxonomy on the genus *Euphrasia* L. (Scrophulariaceae—Rhinanthoideae). Ph.D. Thesis, University of Leicester.

Yeo, P. F. (1962). A study of variation in *Euphrasia* by means of outdoor cultivation. *Watsonia* 5, 224–235.

Yeo, P. F. (1964). The growth of *Euphrasia* in cultivation. *Watsonia* 6, 1–24.

Yeo, P. F. (1966). The breeding relationships of some European *Euphrasiae. Watsonia* 6, 216–245.

Yeo, P. F. (1968). The evolutionary significance of the speciation of *Euphrasia* in Europe. *Evolution* 22, 736–747.

Yeo, P. F. (1970). New chromosome counts in *Euphrasia. Candollea* 25, 21–24.

Yeo, P. F. (1972). *Euphrasia In* "Flora Europaea" (T. G. Tutin *et al.*, eds), Vol 3, pp. 257–266. Cambridge University Press, Cambridge.

Yeo, P. F. (1975). *Euphrasia* L. *In* "Hybridization and the Flora of the British Isles" (C. A. Stace, ed.) pp. 373–381. Academic Press, London and New York.

Yeo, P. F. (1976). Artificial hybrids between some European diploid species of *Euphrasia. Watsonia* 11, 131–135.

Chapter 9

The *Hippuris* Syndrome

C. D. K. COOK

INTRODUCTION

Hippuris is an amphibious plant with simple, linear leaves arranged in symmetrical whorls at regular intervals along the stems (Fig. 1C). The stems are creeping and branched below and usually erect and unbranched above. At any particular node the leaves are equally spaced around the stem. The number of leaves at each whorl and the length, shape and texture of each leaf are variable. This variation is to a large extent influenced by environmental factors. Details of this variation and its control are given by McCully and Dale (1961a, b). *Hippuris* usually has between 12 and 6 leaves at each whorl although extremes of 16, 4 or even 2 are recorded. Generally speaking, stems submerged in water develop more leaves at each whorl than aerial shoots or buried rhizomes. Terrestrially developed leaves are darker green, shorter and thicker than submerged assimilating ones. The flowers of *Hippuris* are relatively small and reduced and are borne singly in the axils of aerial leaves. Pedicels are usually very short or the flowers are sessile.

This character-complex of erect, unbranched stems with simple, elongate leaves borne in symmetrical whorls (heterophylly manifesting itself as variation in number of leaves in each whorl and in individual leaf shape and size) is found in several unrelated plant groups. I intend to call this character-complex the *Hippuris* syndrome.

SYSTEMATIC DISTRIBUTION OF THE
HIPPURIS SYNDROME

Dicotyledons

Hippuridaceae

There is no doubt that the Hippuridaceae belong to the dicotyledons but, like so many other florally reduced and vegetatively specialized aquatic families, the relationships of this group are disputed. In most older works it is placed near or in the Haloragaceae, but various authors have suggested affinities with the Elatinaceae, Lythraceae or Primulaceae. However, embryological and recent phytochemical investigations summarized by Wagenitz (1975) indicate that the Hippuridaceae are allied to the Tubiflorae. The single, inferior carpel of the Hippuridaceae is not otherwise found in the Tubiflorae although some families in this group (Globulariaceae, Phrymaceae and Verbenaceae) occasionally have single carpels and, by contrast, the Gesneriaceae occasionally have an inferior ovary. Because of the ovary structure of the Hippuridaceae some authors have suggested an alliance with the Cornaceae, which are perhaps also related to the Tubiflorae.

Hippuris is found in shallow water in a wide range of habitats throughout the temperate and cold regions of the northern hemisphere. Some ecologically specialized races are found in the Arctic and the Baltic Sea but, as there is no clear cut or discontinuous morphological differentiation, it is usual to recognize a single species, *Hippuris vulgaris* L.

Elatinaceae

The Elatinaceae are cosmopolitan and comprise two genera and about 32 species. The affinities of the family are not clear but it is probably related to the Centrospermae; in many ways it resembles the Frankeniaceae. A close affinity to the Hippuridaceae is most unlikely. The floral morphology of the Elatinaceae is very different from that of the Hippuridaceae. Within the Elatinaceae one species (*Elatine alsinastrum* L.) shows the *Hippuris* syndrome. All other species have lanceolate to orbicular leaves arranged in opposite pairs; the family is well described by Pottier (1927).

The vegetative similarity between *Hippuris vulgaris* and *Elatine alsinastrum* is really amazing. In the non-flowering state it is often difficult to distinguish the two species. The best distinguishing characters in the field are the lack of a well developed rhizome and the somewhat paler green colour of *E. alsinastrum*. It occasionally grows with *Hippuris* but shows a preference for habitats that dry out in summer. In permanent water *E. alsinastrum* can persist as a perennial but it is usually annual. It is found in most of Europe from northern France, S.W. Finland and northern Russia southwards to Algeria. In recent years it has shown a decline and has become extinct in many localities.

Haloragaceae (Haloragidaceae)
The Haloragaceae (excluding Callitrichaceae, Gunneraceae and Hippuridaceae) contain seven genera, of which five contain aquatic species. The family is cosmopolitan but the maximum diversity is found in Australia. The affinities of the Haloragaceae are not clear (Wagenitz, 1975), but the family is perhaps best placed in the Myrtales. A close affinity to the Hippuridaceae is most unlikely. Within the Haloragaceae, whorled leaves are found in the genus *Myriophyllum*. However, not all species have whorled leaves and among those that do most have pinnately divided leaves and are thus not strictly *Hippuris*-like. *Myriophyllum elatinoides* Gaudich. from South America, Tasmania and New Zealand, *M. hippuroides* Nutt. ex Torrey & Gray from western North America and *M. pinnatum* (Watt.) B. S. P. Green from North America and Cuba usually develop pinnately divided leaves underwater, but when they grow in bogs or swamps they frequently develop simple leaves and look very much like *Hippuris*. The specific epithets *"elatinoides"* and *"hippuroides"* are by no means misplaced.

The Indian species *M. oliganthum* (Wight & Arn.) F. Mueller develops simple leaves which are usually spirally arranged at the base of the stem but whorled above. Mature plants resemble *Hippuris*. It must be stressed that the *Hippuris* syndrome in *Myriophyllum* manifests itself only in mature emergent plants, the juvenile and submerged plants do not show this character-complex. In *Hippuris* and *Elatine alsinastrum* the *Hippuris* syndrome is seen throughout the vegetative and generative phases of growth, while in *Myriophyllum* it can be regarded as a modifiable phenotypic state which is not found on all mature plants.

Lamiaceae (Labiatae)
The Lamiaceae comprise about 180 genera, of which five contain aquatic species. A close affinity between the Lamiaceae and the Hippuridaceae or Elatinaceae is most unlikely. Among the aquatic genera *Eusteralis* (*Dysophylla*) from Asia and Australia and *Pogogyne* from western North America are species which show the *Hippuris* syndrome. I have not cultivated or studied any species of these genera in the field so I am unable to describe their ecology and phenotypic variation.

Lobeliaceae (*often included within the Campanulaceae*)
The Lobeliaceae comprise about 30 genera, of which 10 contain aquatics. The monotypic genus *Howellia* from western North America (from Oregon to British Columbia) occasionally shows the *Hippuris* syndrome. I have no first hand knowledge of its ecology or phenotypic flexibility but it occasionally grows together with *Pogogyne* (see Lamiaceae).

Lythraceae
The Lythraceae comprise about 25 genera, of which seven contain aquatic species. The genera *Decodon* and *Rotala* have species with whorled leaves. *Decodon* is monotypic and occurs in southern and eastern North America; fossil material is known from Europe and Asia. It is a large, semi-woody plant with buoyant, arching stems and, although it has whorled leaves, it does not grow like or resemble *Hippuris*. The genus *Rotala* has about 60 species, the majority have leaves in opposite pairs or whorls of three; however, a few species mimic *Hippuris* to a remarkable degree. The following species show the *Hippuris* syndrome: *Rotala hippuris* Makino from Japan (Honshu, Kyushu) (see Fig. 1A); the pantropical *R. mexicana* Chem. & Schlect. (including *R. diglossandra* Koehne, *R. pusilla* Tul. and *R. pygmaea* (Kurz) Rajagopal & Ramayya); *R. myriophylloides* Welw. ex Hiern (including *R. hutchinsoniana* A. Fernandes, *R. longicaulis* A. Fernandes & Diniz, *R. longistyla* L. S. Gibbs, *R. nashii* A. Fernandes and *R. pearsoniana* A. Fernandes & Diniz) from most of tropical Africa but absent in West Africa; *R. verticillaris* L. from S.E. India and Sri Lanka but now possibly extinct; and *R. wallichii* (Hook. f.) Koehne from S.E. Asia (from Burma to Hong Kong and Singapore).
     Rotala hippuris, *R. verticillaris* and *R. wallichii* are allopatric Asian

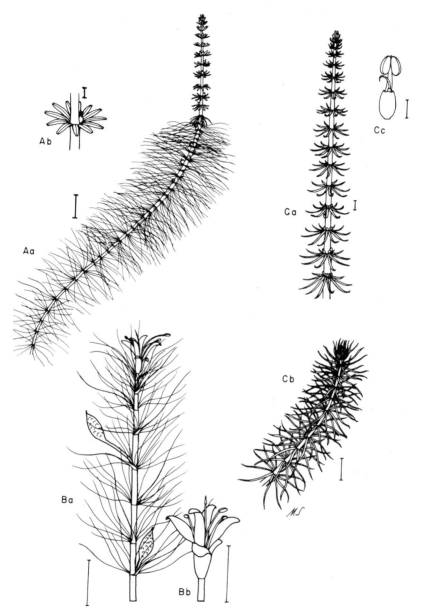

FIG. 1. A. *Rotala hippuris* (Lythraceae): (a) habit, scale 1 cm; (b) flowering whorl, scale 1 mm.
B. *Hydrothrix gardneri* (Pontederiaceae): (a) habit, scale 1 cm; (b) flower, scale 1 cm.
C. *Hippuris vulgaris* (Hippuridaceae): (a) aerial shoot, scale 1 cm; (b) submerged shoot, scale 1 cm; (c) flower, scale 1 mm.

species that closely mimic *Hippuris* throughout their generative history. *R. mexicana* and *R. myriophylloides* mimic *Hippuris* when they are growing as emergent aquatics, but when they are grown terrestrially they creep on the substrate and develop few leaves in each whorl; occasionally they develop leaves in opposite pairs. These terrestrial plants do not resemble *Hippuris* and they have frequently been described as distinct species. This taxonomic confusion is reflected in the number of synonyms cited. In the genus *Rotala* some species show the *Hippuris* syndrome as a modifiable phenotypic state while in other species it is genotypically fixed.

Ranunculaceae
The Ranunculaceae are a large family with about 50 genera and 2000 species. The family is not related to any of the other already listed "*Hippuris* syndrome" families. In this large family only *Ranunculus polyphyllus* Waldst. & Kit. ex Willd. has whorled leaves and shows the *Hippuris* syndrome. The behaviour of this species is well described and illustrated in Glück (1924). Like *Myriophyllum oliganthum* and *Rotala myriophylloides*, it resembles *Hippuris* only when it grows as an emergent aquatic; the terrestrial state has little in common with *Hippuris*. It is a rather rare species found in a few localities in eastern Europe.

Rubiaceae
The Rubiaceae form one of the largest families of flowering plants, with about 6000 species. Many genera have species with whorled leaves. As the majority of aquatic plants with whorled leaves are found in families or orders that normally lack them in terrestrial species, one tends to conclude that this habit is particularly suitable for the aquatic environment. Because the Rubiaceae have numerous terrestrial species with whorled leaves one might expect it to be pre-adapted for "*Hippuris* syndrome" aquatics, but this is not the case. In the whole family only *Limnosipanea spruceana* Hook. fil. (Fig. 2) from Amazonas, Brazil, is aquatic and shows an *Hippuris*-like habit.

Scrophulariaceae
The Scrophulariaceae form a large family with about 220 genera and 3000 species. About 17 genera have aquatic species. The

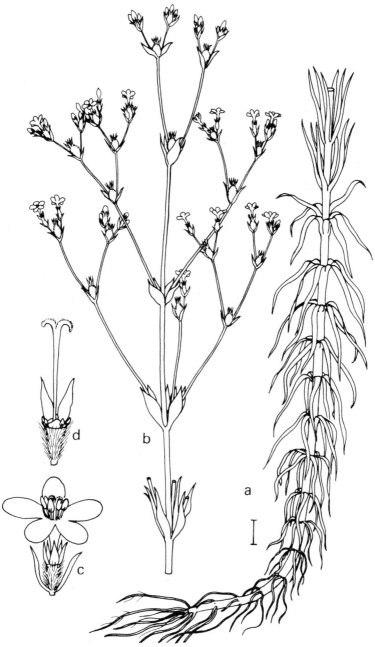

FIG. 2. *Limnosipanea spruceana* (Rubiaceae): (a) submerged shoot, scale 1 cm; (b) inflorescence; (c) flower; (d) young fruit.

Hippuridaceae may well be related to the Scrophulariaceae. Nevertheless, whorled leaves are rare in the Scrophulariaceae but do occur in some aquatic species in the genera *Bacopa* (including *Herpestes*) section Chaetodiscus, *Hydrotriche*, *Limnophila* and *Hemianthus*. However, only two species, *Hydrotriche hottoniiflora* Zucc. from Malagasy and *Limnophila hippuridoides* Philcox from Malaya, can be said to really show the *Hippuris* syndrome. In both species the number of leaves in each whorl is higher on submerged stems, and the leaves themselves are always simple and elongate.

Droseraceae

The Droseraceae are principally terrestrial plants but include one aquatic species, *Aldrovanda vesiculosa* L. *Aldrovanda* does not resemble *Hippuris* and should not be included in the *Hippuris* syndrome but its leaves are in whorls and it is aquatic. However, the leaves bear specialized animal catching traps which function only under water. *Aldrovanda* never develops erect aerial stems.

Monocotyledons

In comparison to the dicotyledonous plants a greater proportion of the monocotyledons are aquatic. However, the *Hippuris* syndrome as I have defined it is not found among the monocotyledons. The nearest approximation is found in the Hydrocharitaceae in the genera *Egeria*, *Elodea* and *Hydrilla*. These are all obligately submerged plants which do not develop aerial stems.

WHORLED LEAVES IN TERRESTRIAL PLANTS

Among terrestrial plants whorled leaves are not particularly rare but they are also not very common. The majority of terrestrial plants with whorled leaves have a regular number of leaves in each whorl. Aquatic plants with the *Hippuris* syndrome have many leaves in each whorl on submerged stems and fewer on aerial stems. This last observation would tend to indicate that a large number of leaves on an aerial stem is not advantageous or, perhaps, that some morphogenetic process prevents large numbers of leaves on aerial shoots. When one considers terrestrial plants, whorls of three or four are common, fives and sixes are rare, and numbers above six very rare. So far in my investigations I have found terrestrial species with consistently more than six leaves in each whorl only in the Rubiaceae. This might be a special case because the majority of "leaves" in the Rubiaceae are generally considered to be stipules. I believe a new and critical investigation into the nature of the "leaves" of the Rubiaceae might well produce unexpected results.

It is possible to make the generalization that terrestrial plants or terrestrial stems of aquatic plants have fewer leaves in each whorl than submerged stems. Because submerged stems have very many more leaves than terrestrial ones there must be some selective advantage in possessing many leaves at each whorl in the submerged environment. The observation that this syndrome has arisen in numerous unrelated aquatic groups also adds considerable weight to this argument.

FUNCTION OF THE *HIPPURIS* SYNDROME

There is general agreement that submerged assimilating organs need and possess a greater surface area than aerial ones; see, for example, Sculthorpe (1967) and Hutchinson (1976). An increase in the number of leaves in each whorl leads to an increase in total assimilating area. This is, however, but one aspect of the *Hippuris* syndrome. The other aspect is that the leaves are in more or less symmetrical tiers. It is more difficult to explain this in physiological terms. Many aquatic plants do not have simple leaves in whorls but by various means have succeeded in developing *Hippuris*-like habits. In many aquatics with whorled leaves the apparently necessary increase in surface area is attained by repeated branching of each leaf and not by increasing the number of leaves. This is shown in diagrammatic form in Fig. 3, where (a) shows the typical *Hippuris* syndrome; (b) is a diagrammatic representation of *Ceratophyllum* in which the leaves are whorled but each leaf is forked; (c) shows whorled leaves but each leaf is divided, as is typical for *Myriophyllum* and *Limnophila*; (d) shows leaves opposite and more divided, as is typical for *Megalodonta* (Compositae) or *Cabomba* (Cabombaceae); and (e) shows single leaves lying in one plane and having an almost circular outline so that they appear tiered, as seen in *Ranunculus circinatus* Sibth. These plants could be regarded, in structural terms, as *Hippuris* mimics. They are very common among aquatics and are brought together in the key based on vegetative characters of aquatics in Cook *et al.* (1974).

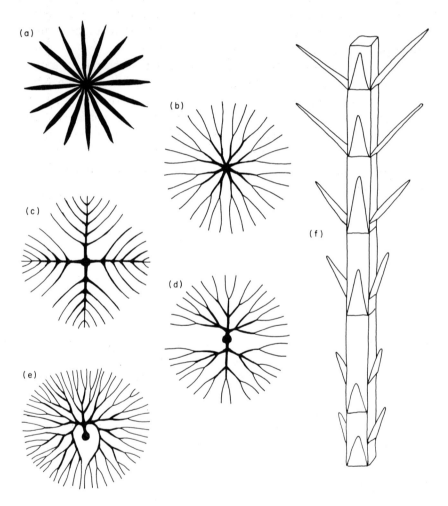

FIG. 3. Diagrammatic representation of the leaf whorl in: (a) typical *Hippuris* syndrome; (b) *Ceratophyllum*; (c) *Myriophyllum* or *Limnophila*; (d) *Cabomba* or *Megalodonta*; (e) *Ranunculus circinatus*; (f) alternate internode elongation in *Rotala densiflora* (Roth ex Roem. & Schultes) Koehne.

THE MORPHOGENESIS OF WHORLED LEAVES

Relatively little work has been carried out on the morphogenesis of whorled leaves. Generally speaking, submerged stems have larger apices than aerial ones (Allsopp, 1964). McCully and Dale (1961a) and Loiseau and Grangeon (1964) demonstrated in *Hippuris* that the larger the circumference of the leaf insertion disc, the larger is the number of leaves. There is little doubt that in *Hippuris* and *Elatine alsinastrum*, the leaves, morphologically speaking, are true leaves. In *Limnosipanea* it is likely that some "leaves" are leaf-like stipules although they have never been critically examined. In *Hydrotriche* all the "leaves" at each whorl arise from two vascular traces. In this case it is likely that the "leaves" are really no more than almost sessile, simple leaflets of two opposite compound leaves; decussate leaves are normal for the Scrophulariaceae. In *Myriophyllum oliganthum* and some species of *Rotala* the juvenile leaves are alternate or in opposite pairs. As the shoot develops some internodes fail to elongate so the leaves become clustered at regular intervals (Fig. 3f). It is clear that no single morphogenetic pathway is responsible for the leaf arrangement in all *Hippuris* syndrome plants. That different morphogenetic pathways lead to structural unity demonstrates the strength of natural selection for a particular growth-form.

Two plants that one can say almost attained the *Hippuris* syndrome are worth mentioning. The first is *Hydrothrix gardneri* Hook. f. (Fig. 1B); it belongs to the Pontederiaceae and, so far, is known only from Ceará, Brazil. At first glance it appears to have whorled leaves, a habit unique in the Pontederiaceae. On closer examination it can be seen that *Hydrothrix* has long shoots with alternate, capillary leaves with sheathing bases; within the sheath there is a short shoot bearing numerous similar but unsheathed leaves. The second is *Carum verticillatum* (L.) Koch (Fig. 4). It is a member of the Umbelliferae and occurs in western Europe from western Spain northwards to Scotland and the Netherlands. In the vegetative phase, when growing in shallow water, it often resembles *Hippuris* in habit. The erect "stems" are in fact leaves; the rachis resembles a stem and the leaflets are arranged in symmetrical tiers and resemble whorled leaves.

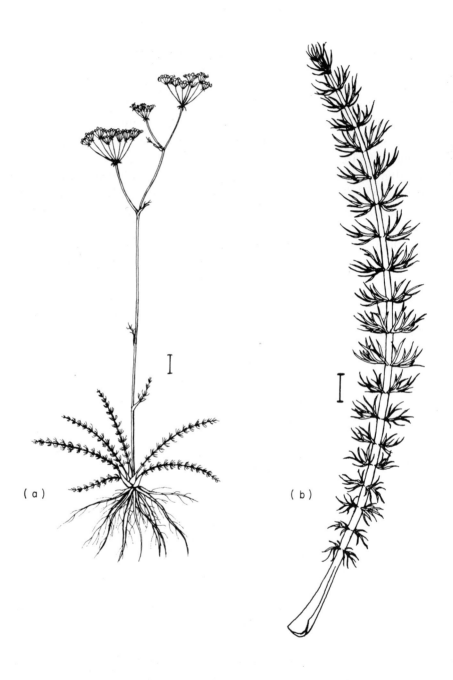

(a)

(b)

FIG. 4. *Carum verticillatum*: (a) habit, scale 5 cm; (b) leaf, scale 1 cm.

DISCUSSION

The *Hippuris* syndrome, when considered to incorporate plants with symmetrical tiers of assimilating organs, is also found among non-flowering aquatic plants, for example *Chara* and other Charophyta, *Draparnaldia* (Chlorophyta), *Batrachospermum* and other Rhodophyta. Among the ferns it is likely that *Equisetum* "acquired" its whorls from aquatic ancestors and it is remarkable that *Salvinia*, the only member of the Filicopsida with consistently whorled leaves, is also aquatic. There can be no doubt that the *Hippuris* syndrome is an example of convergent evolution because it is found in unrelated groups. The structural unity of the syndrome is achieved by different morphogenetic pathways and is therefore not ontogenetically constant, and sometimes there is a tranference of function (the assimilating whorls may be leaves, leaflets, stipules or thalli as in the Algae).

Convergent evolution in response to particular pollination mechanisms or habitat factors such as aridity can be convincingly interpreted in functional terms. However, the reason why symmetrical tiers of assimilating organs are so widespread in aquatic and amphibious plants remains, at least to me, a mystery. Perhaps it is wrong to search for a single factor. Perhaps the syndrome is a result of the interaction of numerous mechanical, physiological, and biotic factors acting together. Nevertheless, the success of this character-complex can hardly be disputed and in botanical teaching it can be used as an example of the complexity of convergent evolution.

REFERENCES

Allsopp, A. (1964). Shoot morphogenesis. *Ann. Rev. Pl. Physiol.* **15**, 225–254.

Cook, C. D. K., Gut, B. J., Rix, E. M., Schneller, J. and Seitz, M. (1974). "Water Plants of the World." W. Junk, The Hague.

Glück, H. (1924). "Biologische und morphologische Untersuchungen über Wasser- und Sumpfgewächse," Vol. 4, Gustav Fischer, Jena.

Hutchinson, G. E. (1976). "A Treatise on Limnology", Limnological Botany, Vol. 3. John Wiley and Sons, New York, London, Sydney and Toronto.

Loiseau, J. E. and Grangeon, D. (1964). Variations phyllotaxiques chez *Ceratophyllum demersum* L. et *Hippuris vulgaris* L. *Mém. Soc. Bot. France* **1963**, 76–91.

McCully, M. E. and Dale, H. M. (1961a). Variations in leaf number in *Hippuris*. A study of whorled phyllotaxis. *Can. J. Bot.* **39**, 611–625.

McCully, M. E. and Dale, H. M. (1961b). Heterophylly in *Hippuris*, a problem in identification. *Can. J. Bot.* **39**, 1099–1116.

Pottier, J. (1927) "Recherches sur l'anatomie comparée des espèces dans la famille des Elatinacées." Imprimerie de l'est, Besançon.

Sculthorpe, C. D. (1967). "The Biology of Aquatic Vascular Plants." Edward Arnold, London.

Wagenitz, G. (1975). Blütenreduktion als ein zentrales Problem der Angiospermen-Systematik. *Bot. Jahrb. Syst.* **96**, 448–470.

Note added in proof

Since going to press I have received living material of *Hydrothrix gardneri* (see p. 173). It is being studied at the moment but it looks as if the "short shoot" theory may need revision. The leaves are arranged almost symmetrically around the nodes and *Hydrothrix* should therefore be considered a *Hippuris* syndrome plant.

Chapter 10

The Taxonomy of Bryophytes

P. W. RICHARDS

INTRODUCTION

The principles of taxonomy are basically the same for all organisms, but the taxonomy of each group has special features. In the case of the mosses and liverworts these depend partly on their unique life-history which normally involves a morphologically and cytologically different gametophyte and sporophyte, each complex enough to provide the taxonomist with a wide range of usable characters. Some taxonomic features of the bryophytes probably also arise from their evolutionary history, which goes back to a very distant past and is far longer than that of the flowering plants. The great antiquity of the group may be partly responsible for, amongst other things, the widespread loss of sexual reproduction (especially among the Hepaticae), the lack of variability ("stenotypy"), and the wide but highly discontinuous distribution of some taxa, characteristics which seem to indicate relict status and evolutionary senescence.

Unfortunately, other features of bryophyte taxonomy do not depend on any inherent characteristics of the plants themselves but simply on the backwardness of the subject. Bryophytes are mostly small and inconspicuous and in a technological civilization they have almost no direct economic importance. The result is that their taxonomy is suffering from years of neglect. This deplorable position will be discussed later; meanwhile it will be enough to say that *Index Muscorum* (Wijk *et al.*, eds, 1959–1969), the equivalent for mosses of *Index Kewensis*, was not completed until 1969*, and *Index*

*A supplement has since appeared (Crosby, 1977).

Hepaticarum is still in progress. The taxonomy of bryophytes of many (especially tropical) countries is nothing less than chaotic, while experimental and other "biosystematic" methods which have been a commonplace for 40 years in the taxonomy of flowering plants have so far had little impact on bryophyte taxonomy.

This essay is no more than a personal view of the present position. For reasons of space it cannot attempt a complete or fully documented survey. It is hoped that it may be useful to those entering the field of bryophyte taxonomy for the first time and also to those whose chief interest lies elsewhere but who wish to understand the position in a field other than their own.

EVOLUTIONARY AND GENETIC BACKGROUND

Age and origin of bryophytes

That bryophytes are a very ancient group is unquestioned though their fossil history is very incomplete. The oldest known bryophyte is the hepatic *Pallavicinites devonicus* (Hueber) Schuster from the Upper Devonian of New York State (Schuster, 1966), but plants which are certainly mosses are not known from earlier than the Permian (Lacey, 1969). In view of the large part that the Anthocerotae have played in speculations on bryophyte phylogeny, it is unfortunate that almost nothing is known of their fossil history. What is more relevant to the taxonomy of modern bryophytes than the age of the whole group or its component classes is that it has been claimed that many, if not most, of the living species of holarctic hepatics are more ancient than existing species of vascular plants. Schuster (1966) even believes that some genera, and perhaps some species, have an age of more than 50 million years. This may well be true of families of hepatics such as the Herbertaceae and Pleuroziaceae in which many species show a loss of sexuality and variability associated with strangely fragmented areas of distribution, as well as of mosses like the small isolated families Schistostegaceae and Bryoxiphiaceae. On the other hand, some groups such as the genera *Frullania* and *Plagiochila* among the hepatics, and moss families like the Bryaceae and Amblystegiaceae, show every sign of active speciation and may be much more recent in origin; many of the species in these groups are extremely variable and in the two families of mosses boundaries between the genera are notoriously difficult to define.

The Bryophyta share a number of characteristics with the green algae (Chlorophyta), the pteridophytes and other green land plants which is a strong indication of a common origin, but modern opinion tends to reject the once prevalent view that they might have evolved directly from some unknown ancient group of green algae. The evidence supporting this theory, such as the apparent similarity of the protonemal phase of mosses to filamentous green algae, has proved to be superficial and misleading. On the other hand, the structure of the gametangia, the cutinized, usually air-borne spores, the stomata in the capsule wall of mosses and anthocerotes, and a good deal of evidence from phytochemistry (Steere, 1969), including the presence of lignins in some mosses (Siegel, 1969)*, link the bryophytes quite closely to the pteridophytes. All these facts lend support to the widely held view that the immediate ancestors of the anthocerotes, and perhaps of all the bryophytes, were simple early psilophytes which had already become adapted to life on land. In addition to the characters the bryophytes share with all the pteridophytes, there is much other evidence, for instance, the remarkable spirally thickened tracheid-like elements in the columella of some anthocerotes (Proskauer, 1960) and the many similarities between the conducting tissues of the Polytrichales and those of the higher plants (Hébant, 1970, 1977), which suggests that the ancestors of the bryophytes had a simple type of vascular system of which only vestiges exist in their modern descendants. The uncanny resemblance between the sculpturing of the spore wall in the living *Dendroceros* and the Devonian *Horneophyton* (Proskauer, 1960) seems to lend plausibility to the notion that the Anthocerotae and the Rhyniales may be at least distantly related. Further evidence on the phylogeny of the bryophytes and their subdivisions may be provided by studies of the amino acid sequence in cytochrome *c* (Boulter *et al.*, 1972).

The theory that bryophytes arose from relatively complex early plants and not directly from algae fits well with the view of Goebel (1930) and others (discussed by Schuster, 1964) that the living bryophytes show more evidence of reduction than of up-grade evolution. In both hepatics and mosses there are excellent examples of morphological series linking different types of organization, for

---

*Hébant (1977) regards the "lignins" of bryophytes as "pseudolignins", i.e. aromatic polymers similar to but not exactly like the lignins of higher plants.

instance the leafy with the thallose Jungermanniales, and *Riccia* with the more complex Marchantiales. In the mosses there are series in several families connecting genera which have simple cleistocarpic capsules (e.g. *Phascum* and *Pleuridium*) with complex genera which have capsules with elaborate peristomes and dehiscence mechanisms. Such series were interpreted by Goebel as illustrating evolutionary trends from complex to simple types of structure: he did not suggest that *existing* simple forms were derived from *existing* complex forms.

It is true that morphological series can often be read either upwards or downwards, but Goebel produced arguments from ontogeny and comparative morphology for believing that in most cases the series showed reduction. Apparently functionless structures such as the vestigial peristome of the cleistocarpic moss, *Pottia bryoides* (Dicks.) Mitt. sometimes provide incontrovertible evidence for reduction. Other evidence for reduction, though possibly open to other interpretations, comes from Burgeff's (1943) genetical work on *Marchantia* in which a series of remarkable mutants came to light, some of which resembled other genera in the same order, such as *Dumortiera, Monoselenium* and *Riccia,* in one character or another. If modern bryophytes are indeed the descendants of more complex land plants, it is not surprising to find evidence of reduction. For as Goebel (1930) has put it, the phylogenetic drama we are trying to understand is only the last act.

Life-history and breeding system

The life-history of mosses and liverworts differs from that of other land plants in being dominated by a gametophyte generation to which the sporophyte is permanently attached and on which it is probably always dependent to some extent for its nutrition. A difference from all other land plants, except the pteridophytes and primitive gymnosperms, is that in bryophytes sexual reproduction depends on the transference of motile spermatozoids, which can survive only in a liquid medium, from the antheridia to the archegonia on the same or another plant.

Though sexual reproduction is usually necessary for spore production, in nearly all bryophytes some method of vegetative multiplication of the gametophyte is found. This can be by gemmae, bulbils on the rhizoids, fragile leaves etc; almost any detached part of the plant can develop into one or more new individuals. Although the vegetative dispersal units of many species seem adapted for dispersal by wind, water or small animals, they often develop close to the parent gametophyte. The majority of liverworts and mosses grow socially in tufts, wefts or carpets which may become several square metres in extent. Colonies are often mixtures of several species, but even when they are of one species there is no easy way of discovering whether they are single individuals, clones or the products of more than one spore. The frequency of sporophytes in some dioecious species, e.g. *Hypnum cupressiforme* Hedw., suggests that the last may often be true and Longton (1974) found evidence that tufts of *Polytrichum alpestre* Hoppe were not genetically homogeneous. A study of this problem using isozymes or radioactively labelled spores might be interesting.

The distance that the spermatozoids have to travel to reach the archegonia may be very short in monoecious (including synoecious and paroecious*) bryophytes, but in dioecious species it depends on whether male and female plants are growing in mixed colonies, close together or widely separated. Spermatozoids move about 100–200 μm s^{-1} and the distance which they can travel by their own movements at normal temperatures is theoretically as much as 1–2 m (J. G. Duckett, personal communication). Probably the actual distance is usually very much less, though as Muggoch and Walton (1942) showed, the spermatocytes are often passively transported for a considerable distance on the surface of water films. In other cases spermatozoids are carried by splash mechanisms. There is little direct evidence of the maximum distance at which fertilization is possible. Bedford's (1938a, b) observations on sporophyte production in the dioecious mosses *Climacium dendroides* (Hedw.) Web. & Mohr and *Breutelia chrysocoma* (Hedw.) Lindb. suggest that in these species it is about 2–7 cm. In *Marchantia* where spermatozoids are transferred by rain splash from the upper surface of the male receptacle to the female receptacles of another plant (at a lower level at the time of fertilization), Burgeff (1943) found that the maximum distance was about 30 cm. Anderson and Lemmon (1974) studied gene-flow distances in *Weissia controversa* Hedw., a

---

*Especially in synoecious species where the antheridia and archegonia are mixed in the same "inflorescence" and paroecious species where the antheridia are immediately below but not mixed with the archegonia.

monoecious moss in which different colonies can be distinguished by different numbers of m-chromosomes. In the habitat studied, bare soil on a roadside bank, transplant experiments showed that the maximum fertilization distance was 40 mm and the median distance was 5 mm. There was no evidence of transfer of spermatozoids from one colony to another.

In sexually reproducing bryophytes gene flow of course depends on the dispersal of spores and other propagules as well as on transfer of spermatozoids. It might be expected that the short fertilization distance would usually limit gene flow to some extent, especially as vegetative growth often leads to the formation of extensive colonies of genetically similar plants. There is little direct evidence on how much outbreeding actually occurs and it is sometimes stated that the sporophytes of monoecious bryophytes are highly homozygous (Schuster, 1966). Gemmell (1950) found that monoecious British mosses are on the whole less widely distributed than dioecious species and suggested that this was because they are more inbred and less variable. On the other hand, Newton (1968b) found that some 7% of the British mosses she examined showed structural heterozygosity. There is also evidence (Lazarenko and Lesnyak, 1972; Ashton and Cove 1972) that some monoecious mosses are self-sterile and therefore obligatorily outbreeding.

About 80% of all hepatics (Schuster, 1966) and over 50% of British mosses are dioecious. In dioecious bryophytes where the two kinds of sex organs are on separate plants, outcrossing is obligatory and the heterozygosity of the spores will generally be greater than in monoecious species, but sexual reproduction is, for obvious reasons, more hazardous. It is not surprising that many dioecious bryophytes rarely produce spores and that some never do so.

In about 14% of British mosses sporophytes have not been found in Britain and in some quite common species they are altogether unknown, e.g. *Oxystegus sinuosus* (Mitt.) Hilp. Many hepatics also are always sterile. Similar proportions of sterile species are found in North America, but according to Schuster (1966) a smaller percentage is found in countries such as New Zealand with a very favourable climate. Some sterile bryophytes produce no archegonia or antheridia at all, but many produce gametangia of only one sex, e.g. the archaic hepatic *Takakia* (Fig. 1), and most are probably potentially dioecious. Behaviour sometimes varies in different places;

thus *Atrichum crispum* (James) Sull. & Lesq., which is widespread and fertile in North America, is always sterile and male in Britain, the only part of Europe where it is found.

Sterile species, as Gemmell (1950) showed, are less variable than both monoecious and fertile dioecious species. Some are "stenotypic" to an extreme degree, combining lack of variability with a very narrow habitat range and a wide but highly discontinuous geographical distribution, e.g. the hepatic *Pleurozia purpurea* (Lightf.) Lindb., populations of which from as far apart as Japan and Scotland are indistinguishable.

It seems then that though selection may favour dioecious breeding systems in bryophytes under some conditions, dioecism may lead to a complete loss of sexual reproduction with a consequent loss of biotypes and ecological versatility.

There is no doubt still much to be learned about the breeding systems of bryophytes, but it is already clear that, as in other plant groups, there are several different systems with different taxonomic implications. In what is probably the commonest system, sexual reproduction is normal and reproduction by spores as well as vegetative reproduction of the gametophyte are effective, but the relative importance of the two methods may vary with ecological conditions. In bryophytes of this type outcrossing is probably frequent, though possibly less effective than in flowering plants, and there is a fair stock of genetic variability on which selection can act. Among such bryophytes taxonomic problems may be not very different from those met with in most groups of flowering plants. But in addition to these evolutionarily vigorous taxa there are many others, especially among hepatics, which through inbreeding and the partial or complete breakdown of sexual reproduction are much less variable. Such taxa are apparently relics, or doomed to become relics. Spores are rarely or never produced and the species tend to be well defined and often rather isolated. In some respects these bryophytes are analogous to the apomictic taxa so common in some families of flowering plants, but unlike them these bryophytes do not seem to abound in "microspecies", differing only slightly from one another. This may be merely because they have not been studied sufficiently or it may be that apomictic higher plants have retained some capacity for crossing and forming new genotypes which sterile bryophytes have entirely lost.

Cytogenetics

General accounts of the cytogenetics of bryophytes have been published by Lewis (1961), Anderson (1964) and Vaarama (1976) and there are several general reviews dealing more specifically with bryophyte chromosomes and chromosome numbers, e.g. Anderson (1974), Berrie (1960) and Steere (1972). Here it will be necessary to deal only with a few general points particularly relevant to taxonomy.

In a moss or liverwort the chromosome number is doubled at syngamy and meiosis takes place at spore formation: thus in normal life-cycles the gametophyte is haploid and the sporophyte diploid, but both generations can be polyploid. For at least 100 years (Pringsheim, 1878) it has been known that apospory in mosses can be induced by cutting off young sporophytes which when placed on a suitable substrate will put out protonemal filaments from which buds and leafy shoots will arise. Apospory can also be induced experimentally by other methods and in some mosses (and a few hepatics) it may occur naturally. Apospory involves the formation of a gametophyte from a sporophyte without the intervention of spores: meiosis is therefore short-circuited and the resulting plants are diploid. In certain races of *Phascum cuspidatum* Hedw. (and perhaps in some other mosses), apospory is linked to apogamy (production of sporophytes without sexual fusion of nuclei), so that as in some ferns the whole life-cycle proceeds without change of chromosome numbers. Wettstein (1928, 1932) produced polyploid series in *Funaria hygrometrica* Hedw. and other mosses by using artificially induced apospory. For bryophyte taxonomy one important conclusion from these experiments is that there is no simple connection between the ploidy of the plant and whether it has a "gametophyte form" or "sporophyte form"; polyploid gametophytes do not usually resemble sporophytes, though gametophytes of different ploidies may differ in cell size and other characteristics.

In bryophytes with a normal life-cycle the same genes are present in both generations but for reasons which are not understood their effects are quite different in sporophyte and gametophyte. In a sporophyte or polyploid gametophyte they may occur in a homo- or a heterozygous form, but not of course in a normal haploid gametophyte.* Both generations must be subject to selection by the environment but the selection pressures on the two generations are probably of different kinds and perhaps act more directly on the gametophyte, which alone is in direct contact with the substratum, than on the sporophyte. This is perhaps why the sporophyte is "evolutionarily conservative" (Steere, 1958) and tends to be alike in different species of the same genus: in some genera the species can be classified on gametophyte characters only. On the other hand, the sporophyte provides most of the important characters for separating orders, families and genera. In mosses it has been recognized since Hedwig's work in the eighteenth century that the peristome above all provides taxonomically useful characters for the higher taxonomic categories. As Heitz (1944) pointed out, the variations in both the major and minor features of moss peristomes are so numerous that it is extremely difficult to imagine what their functional importance can be or to understand how they could have arisen by natural selection.

Two other aspects of bryophyte cytogenetics which are important for taxonomy are hybridization and polyploidy. Schuster (1966) states that no natural hybrids have been found in the Hepaticae "except probably in *Pellia*", but there are numerous reports of naturally occurring moss capsules of supposed hybrid origin. Some of the best attested examples are in the closely related genera *Astomum* and *Weissia*. Nicholson (1905, 1906) recorded several putative natural hybrids in these genera from Sussex, and a rather more doubtful hybrid between *Astomum crispum* (Hedw.) Hampe and *Tortella flavovirens* (Bruch) Broth. (Nicholson, 1910). Anderson and Lemmon (1972) have carefully investigated crosses found in North America between *Astomum ludovicianum* Sull. and *Weissia controversa* (Hedw.) and between the latter and *Astomum muhlenbergianum* (Sw.) Grout. In these hybrids meiosis was regular but the spores were few and of low viability; no gametophytes were obtained from them. It was concluded that the high sterility of the hybrids and the absence of introgression implied effective isolation barriers between the species involved. The occurrence of hybrids perhaps also casts some doubt on the validity of separating *Astomum* from *Weissia*.

Many artificial interspecific and intergeneric crosses were made by Wettstein (1928, 1932) in his work on the Funariaceae. A useful summary of this work is given by Anderson (1964) who discusses its

*But see comment by Longton (1974: 62).

bearing on taxonomy. Burgeff's (1943) work on *Marchantia*, referred to already in another context, includes an account of numerous hybridization experiments between various tropical and European *Marchantia* species. Among the latter was a wide range of material of *M. polymorpha* L., including the varieties *aquatica* Nees and *alpestris* Nees. It is interesting that the varieties *aquatica* and *alpestris* were not very fertile when crossed and crosses of both with typical *M. polymorpha* were unsuccessful. The breeding behaviour as well as the different geographical and ecological ranges of the three taxa clearly indicate that they should be regarded as good species.

Research on chromosomes in bryophytes developed later than in flowering plants, partly because of technical problems; the chromosomes are often very small, especially in mosses. Few reliable counts were available before about 1930. Since then the development of squash techniques has overcome many of the difficulties, but there is still no easy way of counting the chromosomes in gametophytes. Counts are now available for over 1500 bryophyte species (A. J. E. Smith, personal communication), though in some, especially tropical, families the chromosomes are still quite unknown.

The information now available contributes something, but perhaps less than might be expected, to bryophyte taxonomy. In some groups the chromosome number is constant, e.g. in all *Sphagnum* species $n = 19 + 2$. In other groups there is not much variation. The great majority of hepatics have $n = 9$ (i.e. $8 + m$), but there are some other numbers; for example, the very primitive *Takakia* has $n = 4$ which is the lowest number known in bryophytes and is believed to be one of the basic numbers in the class. Families which on other evidence seem to be actively evolving, such as the Bryaceae and Funariaceae, have a wider range of numbers, and in general higher numbers, than families which are more stenotypic and supposedly more ancient.

At family and genus level sweeping generalizations are not possible but chromosome data usually support classifications mainly based on morphology. For example, Lowry's (1948) chromosome counts for *Mnium* (*sensu lato*) agree well with the sub-division of the genus based on other criteria. One argument supporting Smith's (1971) proposal to merge the Dawsoniaceae and Polytrichaceae is that in the majority of species of both families $n = 7$ (though some polyploids are found). On the other hand, the number ($n = 5$) and size of the chromosomes in the moss *Pleurozium schreberi* (Brid.) Mitt. shows that it is out of place in the Entodontaceae (which has consistently $n = 11$), where taxonomists have often placed it.

With species also there is no invariable agreement between chromosomes and taxonomy. Both aneuploidy and polyploidy are found in bryophytes but differences in chromosome number between different populations of the same species are not always linked to differences in structural features or ecology. The taxonomic treatment in such cases is often debatable and certainly cannot be decided by simple rules. Thus some species with a wide geographical and ecological range but no marked morphological differences between populations may have several different chromosome numbers, e.g. *Atrichum undulatum* (Hedw.) PB. The different chromosome "races" resemble species in not being fully fertile when crossed with one another but cannot be separated by normal taxonomic methods. *Riccardia pinguis* (L.) Gray, a variable and widely distributed hepatic, has much variation in its chromosomes, especially in the amount of chromatin, but this does not appear to be correlated with differences in morphology or ecology (Schuster, 1966). In the common moss *Tortula muralis* Hedw. many different chromosome numbers have been found but they do not seem to be correlated with characters of taxonomic importance (Newton, 1968a).

Among bryophytes many examples are known of closely related pairs of species in one of which the gametophyte is haploid and in the other diploid, e.g. *Metzgeria furcata* (L.) Dum. ($n = 8$) and *M. conjugata* Lindb. ($n = 17$) and several pairs of *Mnium* species (Lowry, 1948). Often, but not always, the haploid species is dioecious and the diploid one monoecious or synoecious, e.g. *Rhizomnium punctatum* (Hedw.) Kop. ($n = 7$) and *R. pseudopunctatum* (B. & S.) Kop. ($n = 13$ or 14). In the examples given the two members of the pair differ in easily recognizable characters, as well as in ecological preferences, "inflorescence" and ploidy, but in some cases there is almost no morphological difference except in "inflorescence", e.g. *Bryum pseudotriquetrum* (Hedw.) Schwaegr. ($n = 10$) and *B. bimum* Brid. ($n = 20$). The diploid "race" of *Pellia epiphylla* (L.) Corda ($n = 18$) is sometimes regarded as a distinct species (*P. borealis* Lorb.) but it is morphologically indistinguishable from the haploid "race" with 9 chromosomes (Paton and Newton, 1967). The consensus of opinion seems to be that polyploids should not be treated as distinct species unless there are recognizable differences of structure as well as of chromosome number. Populations which differ in chromosome characters other than number should be similarly treated. Thus Lorbeer (1934) found that European and American plants of *Sphaerocarpus texanus* Aust. differed in the size and shape of the X-chromosome though they were otherwise indistinguishable: the suggestion that the European populations should be separated under the name *S. europaeus* Lorb. has not been generally adopted.

Genecological differentiation and plasticity

Bryophytes which are widely distributed and reproduce by spores (in contrast to the sterile species discussed earlier) are often very variable and occur in many different habitats. The extremely polymorphic moss *Hypnum cupressiforme* Hedw. (which may indeed be a species-complex) is an outstanding example, but other mosses and a few hepatics are almost as variable. Field observations suggest that such species may be plastic and also genecologically differentiated like some higher plants. How far the variation is phenotypic and how far genetically determined can only be definitely decided by experiments.

The pioneer in this field was H. Buch who early in this century began experiments on *Scapania* and other genera of liverworts (Buch, 1922, 1928, 1929, 1933). Buch used simple methods of cultivation and also transplanted material from one natural habitat to another. The object of his work was to show whether the differences between populations were maintained when they were grown together under similar conditions and whether their characteristics changed when they were moved from one habitat to another. He was able to show that some supposed species were only habitat "modifications" (phenotypic variations) of other species, for instance *Scapania dentata* Dum. and *S. intermedia* (Husn.) Pears. were shown to be modifications of the widespread and variable *S. undulata* (L.) Dum. But in his work on the *ventricosa* group of *Lophozia* he found that some populations remained distinct when grown together and in this way he was able to show that the group included a previously overlooked species *L. silvicola* Buch. He also found that many species showed parallel phenotypic variations when grown under similar habitat conditions and suggested a standard descriptive nomenclature for them, e.g. mod. (modification) *pachyderma* for forms with thickened cell walls found under conditions of high evaporation, mod. *colorata* for forms with strongly pigmented walls from well illuminated habitats. A combined name can be used for forms combining two or more characters, e.g. mod. *pachyderma-denticulata-colorata*.

It is unfortunate that Buch has had so few followers; even now, after over 50 years very little work even of the simple kind he proposed has been done on either liverworts or mosses. Some preliminary experiments by the writer (Richards, 1945) showed that the sand-dune moss *Tortula ruralis* subsp. *ruraliformis* (Besch.) Dix. retained its characteristic leaf shape when grown alongside *T. ruralis* (Hedw.) Gaertn., which is mainly a plant of roofs and walls, indicating that it should probably be regarded as an ecotype. Other experiments showed that some varieties of *Hypnum cupressiforme*, e.g. var. *lacunosum* Brid., remained distinct when grown together under similar (partly controlled) conditions (P. Chamberlain, unpublished).

More thorough studies were done on the very variable mosses *Drepanocladus exannulatus* (B. & S.) Warnst. and *D. fluitans* (Hedw.) Warnst. by Lodge (1964) and on the Cuspidata section of *Sphagnum* by Agnew (1958). Further experimental work on *Sphagnum* would be valuable because even simple field observations suggest that some species are extremely plastic. In the past Warnstorf and others gave taxonomic status to every variant whether presumed to be genetically based or not so that innumerable varieties, forms and even subforms were described.

More recently sophisticated experimental methods have become available which could reveal something of the genecological variation underlying the vast amount of phenotypic variation found in some bryophytes; an example is Hatcher's (1967) biometric work on the common liverwort *Lophocolea heterophylla* (Schrad.) Dum. Forman's (1964) work on *Tetraphis pellucida* Hedw. is probably the most detailed study yet made of the responses of a bryophyte to experimentally controlled conditions. *T. pellucida* is not a particularly variable moss and has no named varieties; Forman's main object was not to study variation but to interpret the distribution of the species by determining its tolerance for various environmental conditions. Nevertheless, his work gives a valuable insight into how leaf size and shape respond to light, humidity and other factors when material from a single colony (presumably a clone) is grown in a phytotron.

Longton (1974) studied material from both hemispheres of the bipolar moss *Polytrichum alpestre*. Field studies showed that with increasing severity of climate there was clinal variation in leaf length and in the mean length, dry weight and leaf number of the annual growth increments, so that colonies from near the poles were more compact in habit than those from lower latitudes. In experiments under controlled conditions all these characters were plastic, but there were differences, particularly in leaf length and stem elongation, which were probably genetically controlled.

Though the evidence is still scanty, it appears that some widely distributed bryophytes show "genetically based, ecologically correlated, variation between populations in morphological and physiological characters", as Longton (1974) puts it, but many species are extremely plastic, perhaps more so than most higher plants. Schuster (1966) believes that in hepatics selection has favoured plasticity because their limited capacity for outbreeding and gene exchange has restricted their ability to adapt by producing new genotypes. If this is so, it would apply equally to mosses with similar breeding systems.

THE LOWER TAXONOMIC CATEGORIES

The genus

Since the early eighteenth century classifications, the number of genera of both mosses and hepatics has steadily increased. Linnaeus in his *Species Plantarum* (1753) grouped all the then known bryophytes into 15 genera: these were crude divisions based on the most obvious morphological characters. In Hedwig's *Species Muscorum* (1801), which has been adopted as the starting point of nomenclature for mosses, there are 35 genera, including *Sphagnum* and *Andreaea*. A century and a half later *Index Muscorum* (Wijk *et al.*, 1959–1969) listed over 750 valid generic names in addition to *Sphagnum* and two genera of Andreaeales*. The situation in the

*A third (*Andreaeobryum*) (Fig. 2) has recently been described from northern Alaska and Canada (Steere and Murray, 1976).

Hepaticae is similar: Schuster (1966) lists about 303 genera of Hepaticae and Anthocerotae. The increase in genera is of course partly due to new discoveries but also to a narrower and more precise genus concept arising from a better understanding of structural and other characters.

The splitting still continues but the older genera have not all been treated alike. *Sphagnum*, and with a few minor "splits", *Fissidens*, still remain intact, while others such as *Lophozia* in the hepatics and *Hypnum* in mosses have been repeatedly subdivided. The last named remained as one very large genus well into the nineteenth century, and though Bridel, Schimper and others proposed various divisions, the more conservative taxonomists were not quick to adopt them. In the earlier years of this century Fleischer, Loeske, Brotherus and others carried the splitting of *Hypnum* still further and what had once been a single genus became divided between several families.

Though most of the genera derived from *Hypnum* are not generally accepted, in recent years there has been a slight reaction against that which some consider an excessively narrow genus concept (compare, for instance, Smith, 1978 with earlier European Floras such as Mönkemeyer, 1927). Increasing subdivision of genera is a necessary, and up to a point a desirable, concomitant of the search for a more natural classification, but when pushed to excess (as in the writer's opinion it has been in some families of higher plants), it tends to destroy the utility of the genus concept.

In the bryophytes, as might be expected, satisfactory limits to genera are hardest to arrive at in families such as the Lejeuneaceae, Bryaceae and Hypnaceae, which are probably the most actively evolving. The narrowness of the genera in these families may to some extent reflect the rate of speciation compared with other groups. It is inevitable that in actively speciating groups generic limits should be narrow and rather arbitrary, but there are disadvantages in a very uneven genus concept and perhaps in the next decade bryologists should pay more attention to reaching uniformity in this respect.

The species

If we accept Turrill's (1938) often quoted concept of alpha and omega taxonomy, the species at one extreme is merely a population of morphologically more or less similar individuals whose breeding behaviour and genetical relations are unknown; at the other it is a population of similar individuals which are believed to interbreed and to be genetically isolated from other populations. As taxonomy progresses the former or "α-species" should in due course become replaced by "ω-species", but it is obvious that because data on their breeding behaviour are not available the vast majority, even of higher plant species, are still at the alpha stage: whether they will prove to be genetically isolated interbreeding populations in the omega sense is at present no more than a supposition. In bryophytes so little experimental taxonomy has been done that we rarely know for certain whether even their specific characters are genetically fixed or not. Nearly all bryophyte species must therefore be regarded as "α-species" and, as Crundwell (1970) has said, dealt with by "classical" taxonomic techniques.

This is true of countries such as Europe and North America where the bryophytes are relatively well known and still more of tropical and other inadequately explored regions. Verdoorn (1934: 39), writing mainly with references to south-east Asian hepatics said, "A newly described species is an hypothesis which in course of time may become a probability". Large numbers of species of both mosses and hepatics from overseas have in fact been described from single or very few specimens and often were not seen in the living state by their describers. Such species are an hypothesis in the sense that their publication implies that the type specimen might eventually be shown to represent a population which, in Huxley's (1940) words, is morphologically distinguishable from other populations, has a geographical distribution area, is self-perpetuating as a group and is likely to be not fully fertile when crossed with members of related groups. Further work in the field, the herbarium, and eventually perhaps in the laboratory, may gradually show that the new species has these "omega" attributes so that the hypothesis will become a probability. On the other hand, it may be found that the supposed new species is no more than an abnormal specimen or that it cannot in practice be separated morphologically from some previously described species.

The number of species of hepatics is commonly estimated at about 10,000 and that of mosses at 14,000–14,500 (Touw, 1974). Although many species no doubt remain to be discovered, experienced taxonomists who have worked on "exotic" floras generally agree that these figures would be considerably reduced by critical revisions. Schuster (1964) believes that the number of "good" species of Hepaticae and Anthocerotae in the world is not over about 5000. Touw (1974), who has much experience of the rich Indo-Malayan moss flora thinks the total number of moss species is about 7000, basing this figure on a comparison between the number of species published and those accepted in 17 genera which have been critically revised.

There are several reasons why so many ill-founded species have been described. The most important is that authors, even today, are often ready to publish new species without comparing them with type specimens or even authentically named material of related species. Original descriptions, especially in the nineteenth and early twentieth centuries, were often brief, vague and lacking in precise measurements; sometimes they were not accompanied by adequate drawings. Identifications based on such descriptions without reference to type material are generally quite unreliable.

A further reason for the creation of "bad" species is the prevalence of the concept of the "geographical species", on which several modern taxonomists have commented. By this is meant the view, once common, that bryophyte species are commonly restricted to one hemisphere, one or two continents or one country, so that when a species is discovered in a locality far distant from where it was previously known it should generally be regarded as new, even if there seem to be no clear morphological differences between specimens from the two localities. As Schuster (1964) has shown, this idea has led to much unnecessary multiplication of hepatic species. It has had a similar effect in moss taxonomy: for example, Greene (1976) says that Müller described 51 out of the 52 species of mosses in the first collection from South Georgia as new, but 15 of them were later found to be already known from elsewhere.

Herzog (1926) showed that the geographical areas of bryophyte taxa in general follow similar patterns to those of higher plants. Some species are endemic in quite small areas, e.g. *Campylopus shawii* Wils. ex Hunt which appears to be confined to localities in south west Ireland and western Scotland. In the Hawaiian islands there are about 15 species of *Plagiochila*, all of which seem to be endemic to the group (Inoue, 1976). Other bryophytes are very widespread: many British species can be found throughout the north temperate zone and some of these also exist in the southern hemisphere and the mountains of the tropics. Sometimes the area is very wide but very disjunct, as in *Scapania nimbosa* Tayl., found in the British Isles, Norway, the Himalayas, Hawaii and China. Some bryophytes are more or less cosmopolitan, e.g. *Funaria hygrometrica* Hedw. and the polymorphic *Ceratodon purpureus* (Hedw.) Brid. agg. Neither of these mosses is found in the lowland tropics; in fact very few bryophytes are common to the temperate regions and the lowland tropics. *Campylopus introflexus* (Hedw.) Brid., which seems to have been introduced into Europe about 40 years ago, perhaps by man, and is still rapidly spreading (Richards and Smith, 1975), is one of these.

Since Herzog's book was written much additional information has accumulated on the geographical distribution of bryophytes; in general it supports his main generalization, but it is now seen that many genera and species are more widely distributed than he supposed. For instance, Herzog (1926) gave several examples of genera found in tropical Africa and tropical America, but it is now known that many species are common to these two areas. The lesson for the taxonomist is, as Schuster (1966: 323) says, that "the student can never assume, when he finds a strikingly distinct taxon in a well investigated area, that this taxon is new to science."

Infraspecific taxa

In the older bryophyte Floras, e.g. Limpricht (1890–1904) and Müller (1905–1916), and in the *Student's Handbooks* of Dixon (1924) and Macvicar (1926), infraspecific categories were freely used; they reached a climax in Warnstorf's *Sphagnologia Universalis* (1911). Since then it has been realized that there is little agreement on how these categories should be defined and modern Floras such as Smith's (1978) use them much more sparingly. The modern attitude (Crundwell, 1970) is that if something new is not worth describing as a species it is not worth describing at all.

Yet a classification with no taxa below the species is incomplete; it cannot adequately reflect the variability of natural populations. Also, for a long time there will be a need for a "probationary" taxon for populations which may or may not prove to deserve specific rank; if not named they are in danger of being overlooked. The great problem is that some bryophytes are so plastic that, as has been seen, the underlying genetic variability is "masked" and can only be detected by lengthy culture experiments. There is a need to reconsider, and if possible to obtain a consensus of agreement, on the use of the variety and the subspecies.

Subspecies have been little used in bryophyte taxonomy. Dixon, one of the few taxonomists to use them widely, says (1924: xxiv) that without them there is "no middle choice" between species and varieties. By this he probably implies that intermediates between his subspecies and the related species are commoner than between two full species. Users of his *Student's Handbook* will testify to the usefulness of subspecies in this sense. But modern taxonomists, especially zoologists, generally use subspecies quite differently to mean slightly differentiated variants of a species occupying distinct habitats or geographical areas which do not normally interbreed with the rest of the species. A few bryologists, e.g. Malta (1926), have used subspecies more or less in this way, though Crundwell (1970) advocates a usage similar to Dixon's.

The utility of the variety in bryophyte taxonomy is more questionable: some would regard it as a category debased beyond recall. If it is to survive, it can only be as a portmanteau category embracing both minor variants which may have a genetic basis and insufficiently investigated variants which may later deserve promotion to subspecies or species. Forms and subforms, if worth retaining at all, can be treated as equivalent to Buch's "modifications", and used for variants which are clearly not heritable.

THE PRESENT STATE OF BRYOPHYTE TAXONOMY

Historical

Bryophyte taxonomy first developed in Europe in the eighteenth century. The first book specifically dealing with the classification of bryophytes was Dillenius's *Historia Muscorum* (1741) which also included algae, lichens, lycopods and a few strays such as *Subularia*, though recognizable descriptions of a few mosses and liverworts had been published in herbals etc. before this. During the nineteenth century the taxonomy of European and North American bryophytes advanced rapidly. Numerous "exotic" species were also described, but the explorations of the time, which did so much to stimulate the taxonomy of higher plants, did much less for bryophyte taxonomy, no doubt because of the lack of an economic motive. Up to well into the present century most bryophytes brought back from "overseas" were collected mainly by missionaries, army officers etc. or by naturalists primarily interested in other plants. These collections were worked on by Mitten (1869), Müller (1905–1916) and others who never saw the plants in the living state (and indeed never set foot outside Europe). Two outstanding exceptions were Spruce who worked in South America from 1849 to 1862, and the artist-bryologist Fleischer who first visited Java in 1899. Both were experienced and highly competent bryologists who collected extensively in the tropics and also worked on their own collections. Spruce's *Hepaticae Amazonicae et Andinae* (1884–1885) had much influence on subsequent work on hepatic taxonomy. Fleischer's epoch-making *Musci der Flora von Buitenzorg* (1900–1922) is much more than a moss flora of Java: it includes a classification of mosses which, with some modifications, was used in Brotherus's volumes in *Die Natürlichen Pflanzenfamilien* (1924–1925), still the only available conspectus of the mosses of the world. It is no accident that Spruce and Fleischer, who contributed so much to the development of bryophyte taxonomy, were both able to study tropical bryophytes in the field.

The relative neglect of bryophytes from the tropics and southern hemisphere still continues. Valentine and Löve (1958) distinguish three phases in the development of regional taxonomy: the exploratory, the systematic and the biosystematic. Bryophyte taxonomy has nowhere yet reached the biosystematic phase, but the "bryologically advanced" countries—most of Europe, North America and Japan—have long been in the systematic phase. Elsewhere only a few countries are beyond the exploratory phase—some parts of the Caribbean area, perhaps Java, the Malay Peninsula, New Zealand and extra-tropical Australia. In the remaining areas the bryophyte flora is imperfectly known and the obstacles to progress are considerable.

Bryophyte taxonomy in the "bryologically advanced" countries

Little need be said about the present state of bryophyte taxonomy in the comparatively small part of the world in which the mosses and liverworts are relatively well known. In these countries exploration is nearing the end. But it is not yet over: 634 moss species were included in Richards and Wallace's *Check-list of British Mosses* (1950), while in Smith's *The Moss Flora of Britain and Ireland* (1978), there are 688. Some of the additions were previously regarded as varieties, but a considerable number were new to science, e.g. *Anoectangium warburgii* Crundw. & Hill. Many were species known elsewhere but not before found in the British Isles. Some of the latter may have been accidental introductions, e.g. *Hyophila stanfordensis* (Steere) Smith & Whitehouse, but others certainly were not, e.g. *Dicranodontium subporodictyon* Broth. (previously known only from the Sikkim Himalayas). While many species have been added to our flora, some long-established species have been eliminated or reduced to varieties, e.g. *Ulota bruchii* Hornsch. Between 1950 and 1978 about 20 species of hepatics were added to the British flora, some like *Fossombronia fimbriata* Paton, new to science.

In continental Europe large areas are probably bryologically as well known as Britain, e.g. Scandinavia and most of western Europe north of the Pyrenees. Some parts of Europe, e.g. the Balkans and the Iberian peninsula, are comparatively poorly known. The discovery of *Oedipodiella australis* (Wag. & Dix.) Dix., a member of a mainly southern hemisphere family of mosses, the Gigaspermaceae, in north-eastern Spain (Potier de la Varde, 1958) is evidence that remarkable additions to the flora of Europe can still be expected.

Though exploration needs to continue, the chief needs of bryophyte taxonomy in Europe are for more coordination, for modern Floras of some countries, and perhaps eventually for a bryophyte Flora of the whole continent. There is also still room for progress in methods of description, using quantitative methods, where appropriate. Bryophyte taxonomy in Europe is now ready for some advance from the systematic to the biosystematic phase of development, and it is to be hoped that the lead given by Buch (1929) more than 40 years ago may at length be followed up.

In America, north of Mexico, the situation is not very different

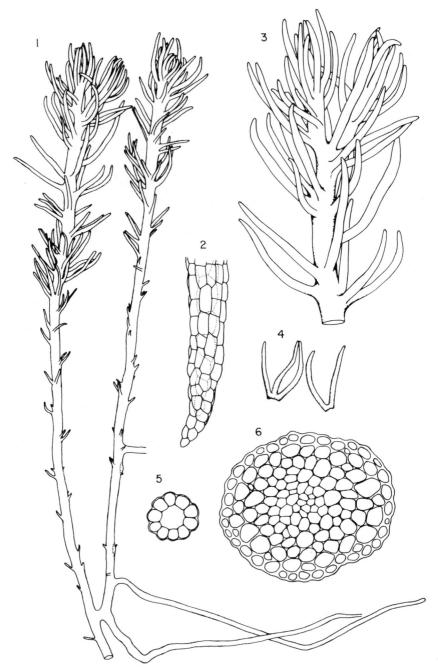

FIG 1. The hepatic *Takakia lepidozioides* Hattori & Inoue. 1. Habit of gametophyte. 2. Apex of leaf lobe. 3. Shoot apex showing terminal branching. 4. Leaves. 5. Leaf lobe in section. 6. Section of stem. After Schuster (1966), reproduced by permission.

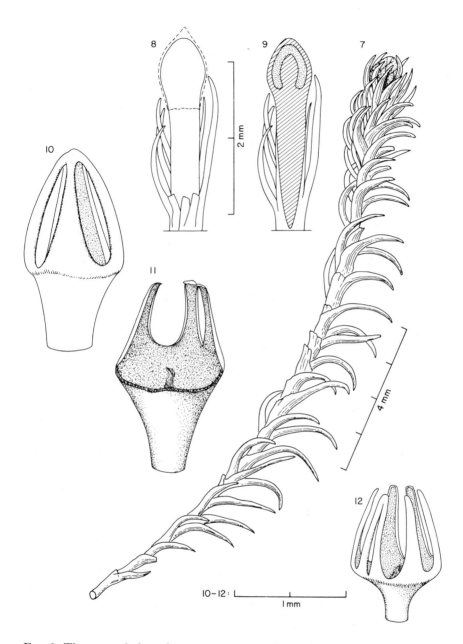

FIG. 2. The moss *Andreaeobryum macrosporum* Steere & B. Murray. 7. Habit of gametophyte. 8. Section of young sporophyte showing calyptra (dashed line). 9. Section of mature sporophyte showing seta. 10. Dehisced capsule with four valves. 11. Old dehisced capsule showing remains of valves and columella. 12. Dehisced capsule with six valves. From original drawings by W. C. Steere, reproduced by permission.

from that in Europe, though it may be true that many parts of the continent are not as well worked as France, Germany or Britain. The part of North America where the most interesting discoveries have been made recently is Alaska. One reason for this is that it is an area of strongly contrasting environments and one in which glacial refuges have allowed relict species to survive (Steere, 1976). One of the most striking discoveries in Alaska is *Andreaeobryum* (Fig. 2) for which Steere and Murray (1976) have created a new family of Andreaeales. No doubt many other interesting species will come to light from this area.

Like Europe, North America has no modern bryophyte Floras covering the whole continent. A hepatic Flora of the United States, east of the Mississippi, is in course of publication (Schuster, 1966) and a moss Flora of the same area is understood to be forthcoming. There is also a modern Flora of the mosses of the Pacific North-West (Lawton, 1971). The surprisingly close relation between the bryophytes of the Appalachian region of the United States and those of eastern Asia has led to increasing cooperation between American and Japanese bryologists to the advantage of both.

Bryophyte taxonomy in other countries

In the tropics and the southern hemisphere, bryophyte taxonomy presents a completely different picture, though of course there are considerable differences from one country to another. Regional Floras of bryophytes, or even check-lists, are few and far between and for most countries the only information available is in papers widely scattered in journals, mostly European, American and Japanese, many of which are old and difficult of access. The chief bryophyte collections from these parts of the world are mainly in European and American herbaria and are as widely scattered as the literature. A special problem is that the herbarium of the nineteenth century German bryologist Carl Müller, who described large numbers of mosses from overseas, was destroyed during the Second World War, so that most of his type specimens no longer exist.

Some reasons for the unsatisfactory state of bryophyte taxonomy in most countries outside Europe, North America and Japan have been briefly mentioned already. The problems involved, for which there can be no quick or easy solution, have been discussed by

Florschütz (1964), Schuster (1964), Touw (1974) and others. Almost everywhere the chief difficulty is that before any real progress can be made, a jungle of inadequately described, poorly understood taxa has to be cleared away; many species—in some countries a large proportion—on critical examination have to be down-graded to varieties or relegated to synonymy. There is nothing to suggest that the species of the southern hemisphere and of the tropics, with the exception of a few notorious "critical" groups, are less well defined than those of the north temperate regions.

The bryophytes of the southern temperate zone which include strange, apparently archaic types such as *Pachyglossa*, *Neohodgsonia* and the moss *Pleurophascum grandiglobum* are better known than those of most tropical countries. A moss Flora of South Africa was written many years ago (Sim, 1926), though a more modern work is no doubt needed now. The plants of Antarctica and the subantarctic islands, including the mosses, have received much attention in recent years by Greene (1976) and others, and can be regarded as relatively well known. New Zealand, for which a Flora of the mosses (Sainsbury, 1955) is available, is perhaps the best known south temperate country bryologically. Australia which until recently had no comprehensive work on either its mosses or its liverworts now has an excellent moss Flora of the southern part of the continent (Scott *et al.*, 1976), but still none for the rich tropical forests of Queensland.

The tropics are not rich in bryophytes everywhere as is sometimes supposed. There is a strong contrast between richness of the mountain forests, even at comparatively low elevations, and the much poorer bryophyte flora of the lowlands. Tropical rain forest at low levels, though enormously rich in higher plants, is relatively poor in bryophytes. A few square kilometres of rain forest in Guyana contained only about 48 species of mosses, a smaller number than in a forested area of the same size in Europe or North America, though the number of hepatics was considerably greater (Richards, 1934, 1954).

In addition to variations in species diversity, there is great variation in the extent to which tropical bryophytes have been collected and studied. The best known tropical areas are probably the Caribbean (including central and parts of northern South America) and the eastern tropics north of the equator. New Guinea and the neighbouring islands, and tropical Africa, are much less well known.

In the Caribbean, Swartz was already collecting mosses in the

H

eighteenth century. Later Spruce's extensive collections from the Amazon and Andes, with others, provided the material for Mitten's *Musci Austro-Americani* (1869) and Spruce's own *Hepaticae Amazonicae et Andinae* (1884–1886). These two great works provided a good foundation for the excellent work of American bryologists in this century.

In a similar way the work of Mitten (1896), Fleischer (1902–1922) and various Dutch taxonomists paved the way for the later work of Dixon (1924), Verdoorn (1934) and others on the mosses and hepatics of the eastern tropics. There is still a great deal of work to be done in this area, both collecting and taxonomic, but it is relatively well known compared with the more southern parts of Indo-Malaya, particularly New Guinea, which appears to be very rich in bryophytes and may yield exciting discoveries in the future.

Africa remains one of the more neglected tropical areas and the situation there deserves a somewhat more detailed discussion. The first bryophytes were sent to Europe by early explorers such as Mungo Park: even the famous Sir Richard Burton collected a few mosses in West Africa in the 1860s and 1880s, but the reputation of tropical Africa as the white man's grave discouraged collecting until well into this century. Africa never had a Spruce or a Fleischer and suffered perhaps more than any other part of the tropics from uncritical taxonomists who described large numbers of bryophyte species from herbarium material only. In recent years the position has begun to improve and the tropical African hepatic flora is much better known, thanks particularly to the patient work of C. Vanden Berghen and E. W. Jones. The position in African moss taxonomy is less satisfactory.

The concept of "geographical species", particularly the failure to realize how closely the tropical African and American bryophyte floras are related, is another source of difficulty. It is now clear that many African mosses are conspecific with, or nearly related to, well known tropical American species.

Tropical Africa is almost certainly less rich in bryophytes than either tropical America or the eastern tropics, as it is in palms and other groups (Richards, 1973), and the number of reputed species will be greatly reduced by critical revision, a slow and laborious process. The effect of revision may be judged from the recent work of Edwards (1976) on *Calymperes*. These mosses have been supposed to be particularly difficult taxonomically: over 40 species have been described from tropical Africa. Edwards has examined type or authentic material of most of these as well as studying the present author's abundant collections and has reduced the number of species from west tropical Africa (as defined by Hutchinson and Dalziel, 1954–1972) to six, of which all but one are identical with American species. *Calymperes* in fact does not seem to be an unusually variable genus in the usual sense, but (except in the section Macrhimanta) the leaves are very polymorphic, some being gemmiferous and some not; two samples from the same tuft may easily appear specifically different. The difficult reputation of the genus largely depends on the brief and vague description of Bescherelle and other earlier authors, which were doubtless often based on inadequate specimens. The variation in leaf shape is only evident when abundant material is available or the plants are studied in the field.

Schultze-Motel has published a useful check-list of West African mosses (1976) and S. R. Edwards and the present writer hope to produce a generic flora of West African mosses in a few years time. But much further work will be needed before the taxonomy of tropical African mosses can be put on a satisfactory basis.

EPILOGUE

At the beginning of this essay the present state of bryophyte taxonomy was described as backward; the following pages will have shown the truth of this statement. Bryophyte taxonomy is certainly backward compared to that of angiosperms or ferns, though not more so than that of various other groups of organisms. The reason is not that bryophytes are a particularly intractable group for the taxonomist, though the practical difficulties of carrying out the sort of biosystematic work which has been done on many angiosperms are rather formidable. Nor are bryophyte taxonomists less competent than others. Undoubtedly the reason is that they are too few, are badly organized and have insufficient support for what they are trying to do.

Taxonomy can be considered an end in itself, like an art, or as a kind of service industry which is necessary for the advancement of other branches of biological science. Amateur and professional taxonomists tend to take the former view but it is for the majority of biologists who probably take the latter view that the backward state of bryophyte taxonomy is a serious problem, because it may hold back the work of ecologists, physiologists and geneticists who are interested in bryophytes from quite different points of view. Yet if we are realistic, we cannot expect that much greater resources will be available for bryophyte taxonomy in the near future or that the number of adequately supported and equipped bryologists will much increase.

Could more be done with not much more than the present resources? One of the greatest problems is that bryologists are scattered all over the world and often work in isolation. If their work was better coordinated, progress could perhaps be more rapid. To achieve such coordination should be one of the objectives of the International Association of Bryologists, founded in 1969. It is not intended to discuss here in detail what might be done, but it might be remarked in conclusion that cooperative projects enthusiastically undertaken are likely to be more fruitful than resolutions and committees. *Flora Europaea* (Tutin *et al.*, 1964–) provides an excellent model: it is providing a most valuable addition to botanical literature, but apart from that the time and money spent have been fully justified by the value of the project in bringing isolated groups of taxonomists together and disclosing unsuspected gaps and inequalities in knowledge. Bryologists could perhaps also find fields for cooperation which would attract, support and enable existing manpower and other resources to be used to better effect. An important field for cooperation might be in the production of critical revisions of difficult groups, which, as many taxonomists have stressed, is one of the greatest needs of contemporary bryophyte taxonomy.

REFERENCES

Agnew, S. (1958). A study in the experimental taxonomy of some British Sphagna (Section Cuspidata) with observations on their ecology. Ph. D. thesis, University of Wales.

Anderson, L. E. (1964). Biosystematic evaluations in the Musci. *Phytomorphology* **14**, 27–51.

Anderson, L. E. (1974). Bryology 1947–1972. *Ann. Missouri Bot. Gard.* **61**, 56–85.

Anderson, L. E. and Lemmon, B. E. (1972). Cytological studies of natural intergeneric hybrids and their parental species in the moss genera *Astomum* and *Weissia. Ann. Missouri Bot. Gard.* **59**, 382–416.

Anderson, L. E. and Lemmon, B. E. (1974). Gene flow distances in the moss, *Weissia controversa* Hedw. *J. Hattori Bot. Lab.* **38**, 67–90.

Ashton, N. W. and Cove D. J. (1976). Auxotrophic and developmental mutants of *Physcomitrella patens. Bull. Brit. Bryol. Soc.* No. 27, 10.

Bedford, T. H. B. (1938a). The fruiting of *Climacium dendroides* W. & M. *Naturalist* **1938**, 189–195.

Bedford, T. H. B. (1938b). Sex distribution in *Climacium dendroides* W. & M. *North Western Naturalist* **13**, 213–221.

Berrie, G. K. (1960). The chromosome numbers of liverworts (Hepaticae and Anthocerotae). *Trans Brit. Bryol. Soc.* **3**, 688–705.

Boulter, D., Ramshaw, J. A. M., Thompson, E. W., Richardson, M. and Brown, R. H. (1972). A phylogeny of higher plants based on amino acid sequences of cytochrome *c* and its biological implications. *Proc. R. Soc. Lond. B*, **181**, 441–455.

Brotherus, V. F. (1924–1925). "Musci (Laubmoose). Die natürlichen Pflanzenfamilien 10–11," 2nd edition. W. Engelmann, Leipzig.

Buch, H. (1922) Die Scapanien Nordeuropas und Sibiriens, 1 and 2. *Teil. Soc. Sci. Fenn. Comm. Biol.* **1**, 1–21.

Buch, H. (1928). Die Scapanien Nordeuropas und Sibiriens, 1 and 2. *Teil. Soc. Sci. Fenn. Comm. Biol.* **3**, 1–177.

Buch, H. (1929). Eine neue Moossystematische Methodik nebst einigen ihrer Resultate und ein neues Nomenklatur system. *Skand. Natur-forskermøde* **18**, 225–229.

Buch, H. (1933). Experimentell-systematische Untersuchungen über die *Lophozia ventricosa* Gruppe. *Ann. Bryol.* **6**, 7–14.

Burgeff, H. (1943). "Genetische Studien an Marchantia." G. Fischer, Jena.

Crosby, M. R. (1977). Index Muscorum Supplementum, 1974–1975. *Taxon* **26**, 285–307.

Crundwell, A. C. (1970). Infraspecific categories in Bryophyta. *Biol. J. Linn. Soc.* **2**, 221–224.

Dillenius, J. J. (1741). "Historia Muscorum." E Theatro Sheldoniano. Oxford.

Dixon, H. N. (1924). "Student's Handbook of British Mosses," 3rd edition. V. T. Sumfield, Eastbourne.

Edwards, S. R. (1976). A taxonomic revision of two families of Tropical African mosses. Ph.D. thesis, University of Wales.

Fleischer, M. (1902–1922). "Musci der Flora von Buitenzorg." 4 vols. E. J. Brill, Leiden.

Florschütz, P. A. (1964). Musci. *In* "Flora of Suriname" (J. Lanjouw, ed.), Vol. 6, Part 1. E. J. Brill, Leiden.

Forman, R. T. T. (1964). Growth under controlled conditions to explain the hierarchical distribution of a moss, *Tetraphis pellucida. Ecol. Monogr.* **34**, 1–25.

Gemmell, A. R. (1950). Studies in the Bryophyta. 1. The influence of sexual mechanism on varietal production and distribution of British Musci. *New Phytol.* **49**, 64–71.

Goebel, K., von (1930). "Organographie der Pflanzen," 3rd edition, 2te Teil, 1 Heft. Bryophyten. G. Fischer, Jena.

Greene, S. W. (1976). Are we satisfied with the rate at which bryophyte taxonomy is developing? *In* Symposium on taxonomy of bryophytes, XII Int. Bot. Congr., Leningrad 1975. *J. Hattori Bot. Lab.* **41**, 1–6.

Hatcher, R. E. (1967) Experimental studies of variation in Hepaticae. I. Induced variation in *Lophocolea heterophylla*. *Brittonia* **19**, 178–201.

Hébant, C. (1970). A new look at the conducting tissue of the mosses (Bryopsida): their structure, distribution and significance. *Phytomorphology* **20**, 390–410.

Hébant, C. (1977). The conducting tissues of bryophytes. *Bryophytorum Bibliotheca* **10**, 1–318.

Hedwig, J. (1801). "Species Muscorum Frondosorum." J. A. Barth, Leipzig.

Heitz, E. (1944). Uber einige Fragen der Artbildung. *Vierter Jahresbericht der Schweizerischen Ges. f. Vererbungsforsch.* **19**, 510–528.

Herzog, T. (1926). "Geographie der Moose." G. Fischer, Jena.

Hutchinson, J. and Dalziel, J. M. (1954–1972). "Flora of West Tropical Africa", 2nd edition, revised by R. W. J. Keay and F. N. Hepper, 3 vols. Crown Agents, London.

Huxley, J. S. (1940). Towards the new systematics, *In* "The New Systematics" (J. S. Huxley, ed.), pp. 1–46. Clarendon Press, Oxford.

Inoue, H. (1976). The concept of genus in the Plagiochilaceae. Symposium on taxonomy of bryophytes, XII Int. Bot. Congr. Leningrad 1975. *J. Hattori Bot. Lab.* **41**, 13–18.

Lacey, W. S. (1969). Fossil bryophytes. *Biol. Rev.* **44**, 189–205.

Lawton, E. (1971). "Moss Flora of the Pacific Northwest." Hattori Botanical Laboratory, Miyazaki.

Lazarenko, A. S. and Lesnyak, E. N. (1972). Comparative study of two moss sibling species—*Desmatodon cernuus* (Hüb.) BSG—*D. ucrainicus* Laz. (Contribution to the problem of infrastructure of the moss species). *J. Gen. Biol.* **33**, 657–666.

Lewis, K. R. (1961). The genetics of bryophytes. *Trans. Brit. Bryol. Soc.* **4**, 11–130.

Limpricht, K. G. (1890–1904). Die Laubmoose Deutschlands, Oesterreichs und der Schweiz. *In* "Rabenhorst's Kryptogamenflora," (L. Rabenhorst, ed.), 3 vols. Eduard Kummer, Leipzig.

Linnaeus, C. (1753). "Species Plantarum." 2 vols, Laurentii Salvii, Stockholm.

Lodge, E. (1964). Studies of variation in British material of *Drepanocladus fluitans* and *Drepanocladus exannulatus*, I and II. *Svensk Bot. Tidskr.* **54**, 368–386 and 387–393.

Longton, R. E. (1974). Genecological differentiation in bryophytes. *J. Hattori Bot. Lab.* **38**, 49–65.

Lorbeer, G. (1934). Die Zytologie der Lebermoose mit besonderer Berücksichtigung allgemeiner Chromosomfragen, I. *Jahrb. Wiss. Bot.* **80**, 565–818.

Lowry, R. J. (1948). A cytotaxonomic study of the genus *Mnium*. *Mem. Torrey Bot. Club.* **20**, 1–42.

Malta, N. (1926). "Die Gattung *Zygodon* Hook. et Tayl." Latvijas Universitates Botaniska Darza Darbi, I. Riga.

Macvicar, S. M. (1926). "The Student's Handbook of British Hepatics," 2nd edition. V. T. Sumfield, Eastbourne.

Mitten, W. (1869) Musci Austro-Americani. *J. Linn. Soc. Bot.* **12**, 1–659.

Mönkemeyer, W. (1927). Die Laubmoose Europas. Andreaeales-Bryales Deutschlands, Oesterreichs und der Schweiz. *In* "Rabenhorst's Kryptogamenflora" Ergänzungsband, Leipzig.

Muggoch, H. and Walton, J. (1942). On the dehiscence of the antheridium and the part played by surface tension in the dispersal of spermatocytes in Bryophyta. *Proc. Roy. Soc. B*, **130**, 448–461.

Müller, K. (1905–1916). Die Lebermoose Deutschlands, Oesterreichs und der Schweiz. *In* "Rabenhorst's Kryptogamenflora," 2nd edition. Eduard Kummer, Leipzig.

Newton, M. E. (1968a). Cyto-taxonomy of *Tortula muralis* Hedw. in Britain. *Trans. Brit. Bryol. Soc.* **5**, 523–535.

Newton, M. E. (1968b). Cytology of British bryophytes. Ph.D. Thesis, University of Wales.

Nicholson, W. E. (1905). Notes on two forms of hybrid *Weissia*. *Rev. Bryol.* **32**, 19–25.

Nicholson, W. E. (1906). *Weissia crispa* × *W. microstoma* C. M. *Rev. Bryol.* **33**, 1–2.

Nicholson, W. E. (1910). A new hybrid moss. *Rev. Bryol.* **37**, 23–24.

Paton, J. and Newton, M. E. (1967). A cytological study of *Pellia epiphylla* (L.) Corda in Britain with reference to the status of *Pellia borealis* Lorbeer. *Trans. Brit. Bryol. Soc.* **5**, 226–231.

Potier de la Varde, R. (1958). Une mousse nouvelle pour l'Europe, *Oedipodiella australis* (Wag. et Dix.) Dix. var. *catalaunica* P. de la V. *Rev. Bryol.* **27**, 11–12.

Pringsheim, N. (1978). Ueber Sprossung der Moosfrüchte und den Generationswechsel der Thallophyten. *Jahrb. Wiss. Bot.* **11**, 1–46.

Proskauer, J. (1960). Studies on Anthocerotales VI. *Phytomorphology* **10**, 1–19.

Richards, P. W. (1934). Musci collected by the Oxford Expedition to British Guiana in 1929. *Kew Bull.* **8**, 317–337.

Richards, P. W. (1954). Notes on the bryophyte communities of lowland tropical Rain forest with special reference to Moraballi Creek, British Guiana. *Vegetatio* **5–6**, 319–328.

Richards, P. W. (1973). Africa, the "odd man out". *In* "Tropical Forest Ecosystems in Africa and South America: a Comparative Review" (B. J. Meggers, E. S. Ayensu and W. D. Duckworth, eds.), pp. 21–26. Smithsonian Institute Press, Washington, D.C.

Richards, P. W. and Smith, A. J. E. (1975). A progress report on *Campylopus introflexus* (Hedw.) Brid. and *C. polytrichoides* De Not. in Britain and Ireland. *J. Bryol.* **8**, 293–298.

Richards, P. W. and Wallace, E. C. (1950). An annotated list of British mosses. *Trans. Brit. Bryol. Soc.* **1**, i–xxxi.

Sainsbury, G. O. K. (1955). A Handbook of the New Zealand Mosses. *Roy. Soc. New Zealand. Bull.* No 5. Wellington.

Schultze-Motel, W. (1975). Beiträge zur Flora von West-Afrika, I. Katalog der Laubmoose von West-Afrika. *Willdenowia* **7**, 473–535.

Schuster, R. M. (1966–). "The Hepaticae and Anthocerotae of North America east of the Hundredth Meridian." 3 vols (in progress). Columbia University Press, New York and London.

Scott, G. M., Stone, I. G. and Rosser, C. (1976). "The Mosses of Southern Australia." Academic Press, London and New York.

Siegel, S. M. (1969). Evidence for the presence of lignin in moss gametophytes. *Am. J. Bot.* **56**, 173–79.

Sim, T. R. (1926). The Bryophyta of South Africa. *Trans. R. Soc. S. Africa* **15**, 1–475.

Smith, A. J. E. (1978). "The Moss Flora of Britain and Ireland." Cambridge University Press, Cambridge.

Smith, G. L. (1971). A conspectus of the genera of Polytrichaceae. *Mem. New York. Bot. Gard.* **21**, 1–69.

Spruce, R. (1884–1885). Hepaticae Amazonicae et Andinae. *Trans. Proc. Bot. Soc. Edinb.* **15**, Part 1, 1–308; Part 2, 309–588.

Steere, W. C. (1958). Evolution and speciation in mosses. *Am. Nat.* **92**, 5–20.

Steere, W. C. (1969). A new look at evolution and phylogeny in bryophytes. *In* "Current Topics in Plant Science." New York.

Steere, W. C. (1972). Chromosome numbers in bryophytes. *J. Hattori Bot. Lab.* **35**, 99–125.

Steere, W. C. (1976). Ecology, phytogeography and floristics of Arctic Alaskan bryophytes. Symposium on Geography and Ecology of Bryophytes. XII. Int. Bot. Congr., Leningrad, 1975. *J. Hattori Bot. Lab.* **41**, 47–72.

Steere, W. C. and Murray, B. M. (1976). *Andreaeobryum macrosporum*, a new genus and species of Musci from northern Alaska and Canada. *Phytologia* **33**, 407–410.

Touw, A. (1974). Some notes on taxonomic and floristic research on exotic mosses. *J. Hattori Bot. Lab.* **38**, 123–128.

Turrill, W. B. (1938). The expansion of taxonomy, with special reference to Spermatophyta. *Biol. Rev.* **13**, 342–373.

Tutin, T. G. *et al.* (eds.) (1964). "Flora Europaea." 5 vols (in progress). Cambridge University Press, Cambridge.

Vaarama, A. (1976). The cytotaxonomic approach to the study of brophytes. *In* Symposium on Taxonomy of Bryophytes. XII Int. Bot. Congr. Leningrad 1975. *J. Hattori Bot. Lab.* **41**, 7–12.

Valentine, D. H. and Löve, A. (1958). Taxonomic and biosystematic categories. *Brittonia* **10**, 153–166.

Verdoorn, F. (1934). Bryologie und Hepaticologie, ihre Methodik und Zukunft. *Ann. Bryol.* Suppl. **4**, 1–39

Warnstorf, C. (1911). Sphagnales-Sphagnaceae (Sphagnologia Universalis). *In* "Das Pflanzenreich" (A. Engler, ed.), pp. 1–546. W. Englemann, Leipzig.

Wettstein, F. von (1928). Morphologie und Physiologie des Formwechsels der Moose auf genetische Grundlage. *Bibl. Genet.* **10**, 1–216.

Wettstein, F. von (1932) Genetik. *In* "Manual of Bryology" (F. Verdoorn, ed.), pp. 232–272. M. Nijhoff, The Hague.

Wijk, W. D. van der, Margadant, W. D. and Florschütz, P. A. (eds) (1959–1969). "Index Muscorum," 5 vols. International Bureau for Plant Taxonomy and Nomenclature, Utrecht.

Chapter 11

The Taxonomy of Lichen-forming Fungi: reflections on some fundamental problems

D. L. HAWKSWORTH

> The concept of lichen is a biological one. 'Lichen systematics' based on algal characters is as unnatural as, e.g. a system of Uredinales based on characteristics from the host plants. (Santesson, 1953: 809)

INTRODUCTION

Germane to any discussion of the systematics of the lichen-forming fungi is a consideration of precisely what is meant by the term "lichenized". The circumscription of an area for taxonomic study in most groups of organisms is on the basis of anatomical and morphological criteria, these features being used in the characterization of the taxonomic groups. In the case of the lichenized fungi, however, it is a single biological feature which essentially delimits the study area. This biological criterion is simply that the fungi regarded as lichenized obtain the carbohydrates they require biotrophically through forming associations with algae; they can perhaps be viewed as phycotrophic fungi (Dobbs, 1970). The nature of the fungus–alga association can be termed "symbiotic" in de Bary's (1879) original sense of unlike organisms living together (Starr, 1975) but whether it is strictly mutualistic is a matter for conjecture. Where a fungus–alga association results in a distinctive and reproducible morphological unit dissimilar from colonies produced by the partners in isolation, the term lichenized is relatively easy to apply. There are, however, other types of fungus–alga relationships, some of which are interpreted as lichenized whilst others are generally passed over by lichenologists. The definition of a lichenized fungus as one exclusively or sometimes living with, but not apparently harming, an alga *and which is studied by lichenologists* has much to commend it as a practical definition of the subject area but can scarcely be accepted as scientific.

In this contribution I propose to first of all discuss this basic typological problem and then to progress to consider (1) some of the difficulties inherent in the production of stable classifications for fungus–alga associations (which largely arise from the nature of the associations themselves), and (2) some nomenclatural problems (which mainly stem from man's attempt to apply rigid rules to very varied situations).

TYPOLOGICAL PROBLEMS

Fungi which are parasites of algae are generally not considered by lichenologists and include some important pathogens of algae, such as the pyrenomycete *Mycophycophila gymnogongri* (Feldm.)A. & J. Cribb, which attacks the marine alga *Grateloupia filicina* (Lamour.)C.Ag., completely destroying it (Cribb and Cribb, 1960). There are approximately 140 fungi known to occur on marine algae, of which about 50 are ascomycetes (Kohlmeyer, 1974; Jones, 1976). Some of these fungi have clear parasitic tendencies similar to those seen in *M. gymnogongri* or occur on decaying algae (i.e. the fungi are saprophytic), but in some instances their relationship with the alga appears to be much more complex. In the case of *Mycosphaerella pelvetiae* Suth., which occurs on the common littoral alga *Pelvetia canaliculata* (L.)Dcne & Thur., the alga is not damaged or modified in any way but at the same time it is extremely difficult to find any specimens of the alga lacking the ascocarps of this fungus; these appear as small black dots and are generally rather regularly scattered over young and older fronds alike. It is consequently evident that *M. pelvetiae* is neither parasitic nor saprophytic and there has been a debate as to whether this fungus–alga association should be termed lichenized (e.g. Smith and Ramsbottom, 1915). A comparable situation occurs in several other marine algae including *M. ascophylli*

Cott. on the alga *Ascophyllum nodosum* (L.)Le Jolis; Kohlmeyer and Kohlmeyer (1972) confirmed earlier reports that the hyphae of *M. ascophylli* ramify widely through the tissues of the host and discussed the possibility that *A. nodosum* and comparable examples should perhaps be considered as lichenized. These authors concluded that as such associations resemble an alga living alone and (recalling Ahmadjian, 1970), ". . . a typical lichen thallus has no resemblance to either a fungus or an alga growing alone" they should not be treated as lichenized; as an alternative the term "mycophycobiosis" was proposed for permanent symbiotic assocations between a marine fungus and a marine alga in which the habit of the alga predominates.

Kohlmeyer and Kohlmeyer's (1972) hypothesis was based on the assumption that the definition of the lichenized state they employed was valid. Unfortunately such a definition fails to embrace the entire range of fungus–alga associations which are traditionally regarded as lichenized; these include both cases where the algal partner predominates and ones in which the fungal partner is dominant. The best example of the former category is afforded by lichens with a filamentous alga as the algal partner (termed "phycobiont") in which the filaments of the alga are simply ensheathed by the hyphae of the fungal partner (termed "mycobiont"), as in *Coenogonium interplexum* Nyl., *Cystocoleus niger* (Huds.)Hariot, *Racodium rupestre* Pers. and *Thermutis velutina* (Ach.)Flot. (see, e.g., illustrations of these taxa in Henssen and Jahns, 1973, and Moser-Rohrhofer, 1975). In the process of lichenization in species of *Collema*, the fungal hyphae appear to ramify through a pre-existing colony of the blue-green alga *Nostoc*; here the presence of the fungus leads to some modification in the shape of the alga but the mycobiont ". . . cannot entirely predominate and determine the form of the lichen" (Degelius, 1954: 30), and the algal cells are not organized into a distinct layer.

In the bulk of the lichenized fungi, the algae become localized in a separate layer sandwiched between layers of fungal tissue (see, e.g., Ozenda, 1963; Henssen and Jahns, 1973; Moser-Rohrhofer, 1975). The impression is consequently one of the fungus being the main determinant of thallus shape (but see pp. 230–232 below); studies of mycobionts in culture support this view for some species (Ahmadjian, 1967a; Werner, 1972). Indeed, it is possible to view the vast array of thallus types developed in the Lecanorales (especially the families Cladoniaceae, Parmeliaceae, Physciaceae, Ramalinaceae,

Stereocaulaceae, Stictaceae and Teloschistaceae) as methods of displaying the enclosed alga to the best advantage of the mycobiont, i.e. in a manner which enables the algal layer to receive the maximum amount of light and hence produce as much carbohydrate for the mycobiont as is possible. This type of adaptation is also seen in the production of dorsiventrally compressed thalli in some characteristically terete species when they occur on the ground instead of hanging from trees (see Hawksworth, 1973, for examples). An analogy may also be drawn between complex lichen thalli and gall formation where host tissue proliferates in a manner beneficial to the gall-inducing agent, in this case the alga (James and Henssen, 1976; see also p. 232).

Intermediate between "typical" lichenized structures in which the association assumes a novel morphological appearance and ones in which the alga is evidently the major determinant of thallus form lies a range of other fungus–alga associations, some of which are extremely difficult to categorize. In Hawaii, for example, Ahmadjian (1970) found that populations of the algae *Scytonema* and *Trentepohlia* showed a progressive envelopment of the algal filaments by fungal hyphae in a more than casual manner; the fungus involved was not determined. Kohlmeyer (1967, 1974) found that *Mycophycophila corallinarum* (Crouan & Crouan)Kohlm., which occurs on at least seven genera of marine algae, is consistently associated with epiphytic algae and, though not generally considered as lichenized, is most satisfactorily viewed as a "primitive lichen". *Herpotrichia juniperi* (Duby)Petr. (syn. *H. nigra* Hartig), causal agent of brown felt blight in coniferous trees, provides another instance of a fungus consistently treated as non-lichenized (Sivanesan, 1972) which has some lichen-like features; Moser-Rohrhofer (1975) has shown that algae can be associated with its hyphae in characteristic nodules. Trees in polluted areas of the British Isles and western Europe generally are frequently colonized by extensive growths of a bright green crust of algal cells generally referred to as *Pleurococcus**; microscopic study of these crusts reveals that the algal cells in it are almost invariably associated with fungal hyphae. Riedl (1976) has ascertained that this "algal" crust is most satisfactorily regarded as

*The nomenclature of this alga is very confused, it is probably most correctly referred to as *Desmococcus vulgaris* Brand (syn. *Protococcus viridis* Ag., *Pleurococcus naegelii* R. Chodat, *P. vulgaris* Nag.); see Bourrelly (1972) for further information.

sterile (i.e. non-ascocarp producing) *Bacidia chlorococca* (Stenh.) Lett.

While there are, consequently, lichen-like associations that perhaps ought to be considered by lichenologists but are generally not, there are also some genera traditionally studied by them in which the species appear to be quite devoid of any true algal partner. As examples the following may be cited: *Arthopyrenia s.s.*, *Leptorhaphis*, *Microthelia*, *Mycoglaena*, *Polyblastiopsis*, *Stenocybe* and *Tromera*; for further information on several of these genera see Harris (1973). In addition there are genera which include both indisputably lichenized and definitely non-lichenized species.

The clearest examples of fungus–alga associations in which the morphological appearance of the fungus is not modified by the presence of an algal partner is perhaps afforded by some of the lichenized basidiomycetes. In *Clavaria corynoides* (Peck)Peters. and *Multiclavula mucida* (Fr.)Peters., for example, small botryose clusters of algal cells enveloped with fungal hyphae occur at the base of the hymenophores (Palm, 1932; Petersen, 1967; Oberwinkler, 1970). Not dissimilar clusters of algal cells are also associated with the agaric genus *Omphalina*. Some bracket-forming basidiomycetes have algae associated within their tissues in a generally well-marked zone (e.g. in *Cora* and *Dictyonema*; see Oberwinkler, 1970) and are readily accepted as lichenized. There are, however, some allied fungi which very commonly have algae on the upper surfaces of their brackets (e.g. *Pseudotrametes gibbosa* (Pers.) Bond. & Sing.) and, while most modern authors would perhaps regard such algae as epiphytic, Wright (1890) considered one sufficiently intimately associated with algae in the British Isles to merit treatment as lichenized.

A further area of difficulty arises within the obligately lichenicolous fungi (i.e. fungi only occurring on lichens). These include fungi which are pathogenic to the lichen causing necrosis and death (e.g. *Lichenoconium erodens* M.S. Christ. & D.Hawksw.; Hawksworth, 1977a), while others induce gall-like deformations (e.g. *Guignardia olivieri* (Vouaux)Sacc.; Hawksworth, 1975). Pathogenic lichenicolous fungi can clearly be regarded as non-lichenized but there are some lichenicolous fungi which co-exist with the host causing no apparent damage (e.g. *Sphinctrina turbinata* (Pers. ex Fr.) de Not.); following Zopf (1897) such associations have been termed "parasymbiotic" (i.e. symbiotic with a pre-existing symbiosis).

Parasymbiotic fungi, as they do not destroy the host tissues, presumably obtain their carbohydrate requirements from the algal partner in the host; consequently, it could be argued that these fungi should be regarded as lichenized. As many lichenicolous species are more closely allied to lichenized than non-lichenized fungal families and genera, there is perhaps a strong case for this approach. James (1965), Santesson (1967) and Hertel (1969), in particular, have noted that a trend in delichenization can be recognized in some genera comprising lichenized and either non-lichenized or lichenicolous species. The following examples of lichen–lichen associations may be viewed as early stages of this trend: *Lecidea insularis* Nyl. with its own discrete thalli on those of *Lecanora rupicola* (L.)Zahlbr. (Hertel, 1970); *Rhizocarpon schedomyces* Haf. & Poelt on a *Pertusaria* species where the *Rhizocarpon* thallus is represented only by a few irregularly organized algal-containing areas (Hafellner and Poelt, 1976); *Acarospora epithallina* Magnusson on *Caloplaca* cf. *paulsenii* (Vain.)Zahlbr. (Poelt and Steiner, 1971), *Arthrorhaphis citrinella* (Ach.)Poelt on *Baeomyces rufus* (Huds.)Rebent. (Santesson, 1960) and *Neonorrlinia trypetheliza* (Nyl.)Syd. on *B.* cf. *pachypus* Nyl. (Poelt and Hafellner, 1976), which are initially dependent on a host lichen but later build up thalli of their own; and *Lecidea insidiosa* Th.Fr., which parasitizes *Lecanora varia* (Hoffm.)Ach., destroying the host plectenchyma and using its algae to produce its own thallus (Poelt, 1974a).

With increasing delichenization (i.e. reduction of the algal-containing thallus) taxa with ascocarps identical to those of typically lichenized genera but lacking any algal partner of their own arise and develop either parasymbiotic or parasitic relationships to their host. These trends are seen in *Arthonia glaucomaria* (Nyl.)Nyl. in apothecia of *Lecanora rupicola* (Hertel, 1969), *Bacidia plumbina* (Anzi)R.Sant. on thalli of *Parmeliella plumbea* (Lightf.)Vain. (Santesson, 1960), *Buellia bayrhofferi* (Schaer.)Oliv. on various brown *Parmelia* species (Keissler, 1930; Santesson, 1967) and *Lecidea associata* Th. Fr. on thalli of *Thamnolia vermicularis* (Sw.) Ach. ex Schaer. *s.l.* (Keissler, 1930), which all form parasymbiotic associations with their hosts. *Buellia destructans* (Tobl.)R.Sant. *ined.* on *Chaenotheca chrysocephala* (Ach.)Th.Fr. and *Catillaria aggregata* (Bagl. & Car.)R.Sant. *ined.* on *Peltigera* species are examples of species parasitizing their hosts (Santesson, 1967). A few genera are known which include lichens, lichens confined to lichens, parasymbionts, fungal parasites of lichens, and saprophytes occurring alone on wood, e.g. *Buellia* (Santesson, 1967).

Not all lichenicolous fungi are allied to clearly lichenized fungi and some belong either to genera which predominantly include plant parasitic and saprophytic fungi (e.g. *Leptosphaeria*, *Microdiplodia*, *Microthyrium*, *Nectria*, *Nectriella*, *Pezizella*, *Phoma* and *Pleospora*), or genera parasitizing non-lichenized ascomycetes, e.g. *Cornutispora* (Hawksworth, 1976a). Other lichenicolous fungi belong to genera comprising only lichenicolous species; some of these have perhaps been derived from lichenized taxa (e.g. *Abrothallus*, *Phacopsis*), whilst others perhaps have their strongest affinities with non-lichenized fungi, e.g. *Clypeococcum* (Hawksworth, 1977b).

The genus *Omphalina* (Agaricales) also includes species with a wide range of nutritional types, comprising: (1) almost certainly lichenized species such as *O. hudsoniana* (Jenn.) Bigelow [with squamules named *Coriscium viride* (Ach.) Vain. (Bigelow, 1970; Henssen and Kowallik, 1976)]; (2) species with *Botrydina vulgaris* Bréb. ex Meneghini algal glomerules around their stipe bases as in *O. ericetorum* (Fr.) M. Lange, *O. griseopallida* (Desm.) Quél. and *O. luteovitellina* (Pilàt & Nannf.) M. Lange (Bigelow, 1970; Hawksworth, 1972); (3) *O. cupulatoides* P. D. Orton, probably parasitic on *Peltigera canina* (L.) Willd. (Orton, 1977); (4) the muscicolous *O. sphagnicola* (Peck) Bigelow (Bigelow, 1970); and (5) saprophytic species such as *O. chrysophylla* (Fr.)Murrill on decaying logs (Bigelow, 1970).

From the preceding discussions it will be evident that there is a very wide range of fungus–alga, fungus–lichen and lichen–lichen associations. These are indicated in Fig. 1 together with suggested evolutionary interrelationships between them; transitional types between many of these categories exist and, within different taxonomic groups, there are indications that there has been evolution in different directions along the same lines. In addition to the complexity of this figure and the infrageneric variations in nutritional methods mentioned above (see also Table III), it is important to recognize that taxonomically remote groups of fungi participate in lichenized associations. These are not restricted to various orders of Ascomycotina (see Table II), which nevertheless constitute most of such associations, but also comprise representatives of the

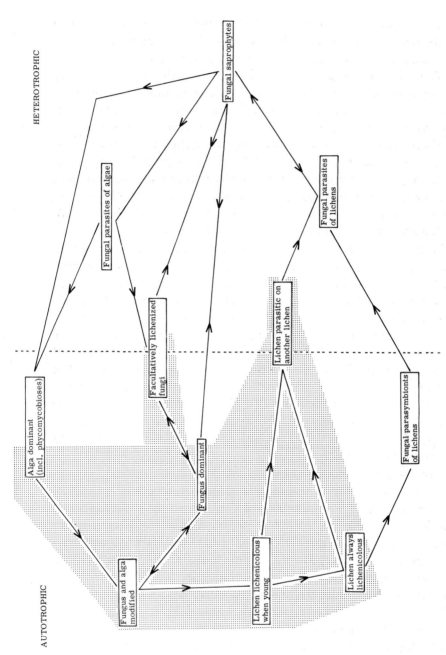

AUTOTROPHIC

HETEROTROPHIC

Alga dominant
(incl. phycomycobioses)

Fungal parasites of algae

Fungal saprophytes

Facultatively lichenized
fungi

Fungus dominant

Fungus and alga
modified

Lichen parasitic on
another lichen

Fungal parasites
of lichens

Lichen lichenicolous
when young

Lichen always
lichenicolous

Fungal parasymbionts
of lichens

FIG. 1. Schematic representation of the types of fungus-alga relationships known. Arrows indicate directions of probable evolution; the stippled area indicates areas including taxa traditionally treated by lichenologists.

Basidiomycotina (Agaricales and Aphyllophorales), Deutero-mycotina (Coelomycetes and Hyphomycetes), and even a "phyco-mycete" (*Geosiphon pyriforme* (Kütz.)Wettst.) Instances are also known of distinct gelatinous thallial structures being formed by Actinomycetes and green algae ("actinolichens") and, at least in culture, myxomycete-green algal associations can be produced ("myxolichens").

These considerations suggest that the lichenized habit has not only developed in diverse groups of fungi at various chronologically and taxonomically remote points during evolutionary history, but is also arising in some groups and being lost in others today. The ability of fungi to form lichen-like associations with algae has often been considered a particularly ancient trait. Cain (1972) suggested that the first heterotrophic ascomycetes might have been lichen parasites, and Kohlmeyer (1975) produced evidence indicating that ascomycetes themselves might have originated from red algae epiphytic on larger seaweeds. The theory that vascular plants might also have arisen from fungus-alga associations, proposed by Church (e.g. Church, 1921), has recently been re-advocated by Pirozynski and Malloch (1975).

In view of the extreme polyphylly of the lichenized habit and the range of fungus–alga relationships encountered in taxa generally regarded as lichens, it will be appreciated that it is extremely difficult to define precisely what is meant by the term lichenized on the basis of the physical relationships of the bionts. There are, however, some other characteristics which might contribute to a more satisfactory definition. James (1965) stressed the perennial ascocarps but, while this holds true for the discocarpous lichens, this is not applicable to all as there are numerous perennial non-lichenized pyrenocarpous fungi. Lichens are almost invariably slow growing (see the review of Hale, 1974a) and within them simple sugars or polyols (depending on the algae involved) pass from the phycobiont to the mycobiont (see the review of carbohydrate transfer in lichens by Richardson, 1974). It has also recently been discovered that a phenomenon termed "physiological buffering" and involving a polyol pool is involved in carbohydrate transfer (Farrar, 1976); this might be indirectly responsible for the slow growth rates encountered by limiting the amounts of photosynthetically fixed carbon available for growth. As rather few fungus–alga associations have been studied physiologically it is difficult to forecast if a definition in metabolic terms might eventually prove possible; because of the similarities in metabolism found in a wide range of biotrophic associations (Smith, 1976), this currently seems unlikely.

In summary, one is forced to conclude that a simple definition of what biotrophic fungus–alga associations should be regarded as lichenized is not possible at the present time. The limits for study by lichenologists are consequently equally poorly demarcated and in practice largely historically rather than scientifically defined. The uncertainty of what constitutes a lichen is of long-standing; Linnaeus (1753), for example, included species now consistently treated as lichenized both in the algal genus *Byssus* and the fungal genus *Mucor*, as well as in *Lichen*.

TAXONOMIC PROBLEMS

Suprageneric concepts

Lichenized and non-lichenized taxa were treated quite separately by all early mycologists and lichenologists and independent systems of subclasses, orders and families devised; that most widely used for lichenized fungi came to be that of Zahlbruckner (1903–1907), in which the major categories were based on the reproductive structures of the mycobiont as they were in non-lichenized fungal systems. Nannfeldt (1932) asserted that lichens were a biological and not a taxonomic group and recognized the fundamental importance of differentiating between ascohymenial and ascolocular types of ascocarp development. Luttrell (1951) discovered that these ontogenetic types were correlated with unitunicate and bitunicate asci, respectively, and included lichenized taxa in an arrangement of all ascomycetes he proposed.

These movements were soon supported by lichenologists (e.g. Santesson, 1952, 1953) and Hale (1957) provided an "... in part speculative ..." disposition of lichen families within Luttrell's system. The arrangement of Hale, with only minor changes, has been widely used (e.g. Duncan, 1970; Galun, 1970), but its vindication has had to await detailed ontogenetic studies on numerous lichenized genera. (The only major departure from this system is that of Kreisel

(1969), who referred the Lecanorales (incl. Thelotremaceae, Buelliaceae), Hysteriales (incl. Arthoniales, Graphideales) and Caliciales all to the Loculoascomycetidae apparently in the belief that their asci were all bitunicate.) The studies, in particular of Professor M. Chadefaud and his students in Paris, culminating in the papers of Chadefaud *et al.* (1969), Letrouit-Galinou (1968) and Janex-Favre (1971), however, began to reveal a much larger amount of variation than had hitherto been suspected. Contemporaneous studies on asci in lichens furthermore revealed that many discocarpous genera, especially in the Lecanorales, had ones intermediate between the basic unitunicate and bitunicate types; these have been termed "arrested bitunicate" or "archaeascé", show considerable ranges of apical structures, and were interpreted as ancestral to simpler unitunicate types (Chadefaud *et al.*, 1969; Letrouit-Galinou, 1973). Only in the last few years have sufficient amounts of ontogenetic data been accumulated to make attempts to integrate lichenized taxa more fully into fungal systems more meaningful.

The incorporation of the perhaps mainly secondarily lichenized basidiomycete lichens presented few problems (e.g. Poelt, 1974b) but that of the probably primarily or at least very anciently lichenized discomycetous lichens proves difficult. As estimates of the numbers of species of fungi (Ainsworth, 1971; see Fig. 2) reveal that over half of all known ascomycetes are lichenized (*c.* 18,000 *vs.* 15,000 species)

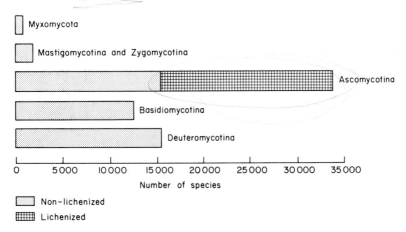

FIG. 2. The numbers of species of fungi in the various divisions of the group. The lichenized taxa of the Basidiomycotina and Deuteromycotina are too few to indicate at this scale. Based on the data of Ainsworth (1971).

TABLE I. *Classification of the lichenized Ascomycotina proposed by Henssen and Jahns (1973).*

	Orders	Suborders	Families
ASCOHYMENIALE PILZE	Caliciales		Caliceaceae, Cypheliaceae, Sphaerophoraceae Ahn. Mycocaliciaceae
	Lecanorales	Lecanorineae	Baeomycetaceae, Acarosporaceae, Arctomiaceae, Candelariaceae, Cladoniaceae, Coccocarpiaceae, Collemataceae, Heppiaceae, Lecanoraceae, Lecideaceae, Pannariaceae, Parmeliaceae, Ramalinaceae, Stereocaulaceae, Umbilicariaceae
		Lichineae	Lichinaceae
		Peltigerineae	Peltigeraceae, Placynthiaceae, Stictaceae
		Teloschistineae	Teloschistaceae
		Physciineae	Physciaceae
		Pertusarineae	Pertusariaceae, Trapeliaceae
	Gyalectales		Gyalectaceae
	Ostropales	Ostropineae	Ostropaceae, Thelotremaceae Ahn. Asterothyriaceae
		Graphidineae	Graphidaceae
	Sphaeriales		Porinaceae Ahn. bitunicate Familien Microglaenaceae, Pyrenulaceae, Strigulaceae
	Verrucariales		Verrucariaceae
ZWISCHENGRUPPE	Arthoniales		Arthoniaceae, Lecanactidaceae, Opegraphaceae, Roccellaceae
ASCOLOCULARE PILZE	Pleosporales		Arthopyreniaceae
	Dothideales		Mycoporaceae Ahn. *Microthelia aterrima*-Gp.

it might have been anticipated that to include so many taxa in schema devised on the basis of under half of the potential variation would not be a process of simple intercalation. In practice most mycologists have tended to be myopic with respect to lichenized taxa (e.g. Korf, 1973), although there have been a few important exceptions (e.g. Luttrell, 1973; von Arx and Müller, 1975). The first ontogenetically based attempt at a comprehensive incorporation of lichenized fungi into the higher categories used for non-lichenized species was that of Henssen and Jahns (1973). Despite a great deal of original research to supplement existing information, these authors were unable to produce an entirely satisfactory scheme and many areas of uncertainty remained (Table I). Being charged with the arrangement of lichenized and non-lichenized ascomycete entries for the next edition of the *Dictionary of the Fungi*, I had begun to consider if a more fundamental revision of ascomycete classification might be necessary when a paper by Barr (1976), a leading pyreno-mycetologist, appeared. Drawing on the information skilfully compiled by Henssen and Jahns (1973) for the lichenized taxa, Barr was able to compile a new system of ascomycete classification within which some lichenized orders and families were correlated more closely with non-lichenized groups than had previously been possible (Table II). Although some parts of this scheme are certainly a matter of debate (e.g. position of the Verrucariales), it has a great deal to recommend it as a framework within which the affinities of particular families and genera can be discussed.

Barr (1976) disposed lichenized families through nine orders in six subclasses. Each of these subclasses also included non-lichenized taxa; something which re-emphasizes the biological rather than taxonomic concept of "lichen" and also their polyphyletic nature. She considered that the Caliciales and Lecanorales, orders within which the majority of lichenized genera fall, to have arisen at an early stage in the evolutionary development of the Elaphomycetidae; this may explain why ascus structures in the Lecanorales particularly are more complex than in other orders of the subclass.

Generic concepts

Dughi (1954), in an important but rarely cited essay on the bases of lichen systematics, emphasized that three distinct sets of characters

TABLE II. *Classification of the Ascomycotina proposed by Barr (1976).*
(Families including lichenized species have been placed in **bold face***)*

Orders	Suborders	Families
HEMIASCOMYCETES	Protomycetales	Protomycetaceae
	Taphrinales	Taphrinaceae
	Spermophthorales	Spermophthoraceae
	Cephaloascales	Cephaloascaceae
	Dipodascales	Dipodascaceae, Hemiascosporiaceae, Eremascaceae
	Ascoideales	Ascoideaceae
EUASCOMYCETES		
PLECTASCOMYCETIDAE	Eurotiales	Amorphothecaceae, Gymnoascaceae, Pseudeurotiaceae, Trichocomaceae, Monascaceae, Eoterfeziaceae, Onygenaceae, Ascosphaeraceae
LABOULBENIOMYCETIDAE	Spathulosporales	Spathulosporaceae
	Laboulbeniales	Ceratomycetaceae, Laboulbeniaceae, Peyritschiellaceae
PARENCHEMYCETIDAE	Erysiphales	Erysiphaceae
	Meliolales	Meliolaceae
	Diaporthales	Gnomoniaceae, Valsaceae, Pseudovalsaceae, Melanconidaceae
	Sordariales	Sordariaceae, Lasiosphaeriaceae, Halosphaeriaceae, Melanosporaceae, Ophiostomataceae
	Coronophorales	Coronophoraceae
ANOTEROMYCETIDAE	Ostropales	**Stictidaceae, Thelotremaceae, Asterothyriaceae, Graphidaceae**
	Gyalectales	**Gyalectaceae**
	Clavicipitales	Clavicipitaceae
	Hypocreales	Hypocreaceae
	Chaetomiales	Chaetomiaceae
ELAPHOMYCETIDAE	Mediolariales	Mediolariaceae
	Cyttariales	Cyttariaceae
	Coryneliales	Coryneliaceae
	Pezizales	Sarcosomataceae, Sarcoscyphaceae, Ascobolaceae, Pezizaceae, Morchellaceae, Helvellaceae, Pyronemataceae

Class	Subclass	Order	Families
		(Helotiales)	Ascocorticiaceae, Hemiphacidiaceae, Geoglossaceae, Sclerotiniaceae, Orbiliaceae, Dermateaceae, **Mycocaliciaceae**
		Lecanorales	**Collemataceae, Parmeliaceae, Lecanoraceae, Lecideaceae, Candelariaceae, Baeomycetaceae, Cladoniaceae, Stereocaulaceae, Umbilicariaceae, Ramalinaceae, Acarosporaceae, Arctomiaceae, Pannariaceae, Coccocarpiaceae, Heppiaceae, Lichinaceae, Placynthiaceae, Peltigeraceae, Stictaceae, Teloschistaceae, Physciaceae, Pertusariaceae, Trapeliaceae**
		Caliciales	**Caliciaceae, Cypheliaceae, Sphaerophoraceae**
		Rhytismatales	Rhytismataceae, Cryptomycetaceae
		Phyllachorales	Phyllachoraceae, Physosporellaceae, Phacideaceae, **Porinaceae**
		Xylariales	Trichosphaeriaceae, Calosphaeriaceae, Amphisphaeriaceae, Diatrypaceae, Coniochaetaceae, Hypoxylaceae, Boliniaceae, Microascaceae
LOCULOASCOMYCETES	LOCULOPLECTASCOMYCETIDAE	Myriangiales	Saccardinulaceae, Atichiaceae, Saccardiaceae, Myriangiaceae, **Arthoniaceae,** Schizothyriaceae, Stephanothecaceae, Leptopeltidaceae
	LOCULOPARENCHEMYCETIDAE	Asterinales	Asterinaceae, Brefeldiellaceae, Englerulaceae, Aulographaceae, Capnodiaceae, Parodiopsidaceae
		Dothideales	Pseudosphaeriaceae, Dothioraceae, **Mycoporaceae,** Euantennariaceae
	LOCULOANOTEROMYCETIDAE	Chaetothyriales	Metacapnodiaceae, Herpotrichiellaceae, Naetrocymbaceae, Chaetothyriaceae, Trichothyriaceae
		Verrucariales	**Verrucariaceae,** Tichotheciaceae
	LOCULOEDAPHOMYCETIDAE	Hysteriales	Parmulariaceae, Phillipsiellaceae, **Opegraphaceae, Patellariaceae, Roccellaceae,** Hysteriaceae
		Pleosporales	**Strigulaceae,** Microthyriaceae, Micropeltidaceae, **Arthopyreniaceae,** Dimeriaceae, Cucurbitariaceae, Phaeotrichaceae, Lophiostomataceae
		Melanommatales	**Trypetheliaceae,** Melanommataceae, Lophiaceae, Zopficaceae

were employed in lichen taxonomy: those of the alga, those of the fungus, and those of the association (consortium).

A difference in algal partners alone is not now accepted as a sole criterion for generic separation and pairs of genera previously segregated on this basis have now been united (e.g. *Lobaria* and *Lobarina*, *Peltidea* and *Peltigera*, *Sticta* and *Stictina*). As will be evident from the discussion of typological problems presented above, even the presence or absence of an algal partner can hardly be maintained as a generic criterion *per se*. Although there have been attempts to apply this dictum in the past (e.g. Vainio, 1922) and more recently (Salisbury, 1976), such a view cannot be deemed as satisfactory as it only serves to isolate what are otherwise indisputably very closely allied taxa on the basis of a single character. Table III lists some examples of genera with lichenized or lichenicolous species

TABLE III. *Examples of genera cutting across biological boundaries. (See text for further explanation.)*

Acarospora Massal.	*Microdiplodia* Allesch.
Arthonia Ach. (incl. *Celidium* Tul., *Conida* Massal.)	*Microglaena* Körb. (incl. *Polyblastiopsis* Zahlbr.)
Arthopyrenia Massal. (incl. *Acrocordia* Massal.)	*Microthyrium* Desm.
Bacidia de Not. (incl. *Mycobilimbia* Rehm)	*Muellerella* Hepp ex Müll. Arg. (incl. *Pleurisperma* Siv.)
Biatorella de Not. (incl. *Tromera* Massal.)	*Nectria* Fr.
Buellia de Not. (incl. *Karschia* auct. p.p.)	*Nectriella* Nits.
Calicium Pers.	*Omphalina* Quél. (incl. *Botrydina* Bréb. ex Meneghini p.p., *Coriscium* Vain.)
Catillaria Massal. (incl. *Scutula* Tul.)	*Pezizella* Fuckel.
Chaenothecopsis Vain.	*Phoma* Sacc.
Coniocybe Ach. (? incl. *Roesleria* Thüm & Pass.)	*Pleospora* Rabenh. ex Ces. & de Not.
Cyphelium Ach.	*Polycoccum* Saut. ex Körb. (incl. *Didymosphaeria* Fuckel, *Lophothelium* Stirt. etc.)
Didymella Sacc.	*Rhizocarpon* Ram. ex DC.
Guignardia Viala & Ravaz (incl. *Laestadia* Auersw.)	*Sphinctrina* Fr.
Lecanactis Eschw. (incl. *Lecanidion* Endl.)	*Stictis* Pers. ex Fr. (incl. *Conotrema* Tuck.)
Lecidea Ach. (incl. *Nesolechia* Massal.)	*Stigmidium* Trevis. (incl. *Pharcidia* Körb.)
Leptosphaeria Ces. & de Not.	*Thelocarpon* Nyl. (incl. *Ahlesia* Fuckel)

which also include species which are either lichenicolous or lichenized, respectively, or have other modes of nutrition.

The features of paramount importance in the characterization of genera including lichenized species are derived from the ascocarps, strictly fungal features, but in practice these are often supplemented by characters derived from vegetative anatomy and morphology or the chemical products of the thallus, attributes largely of the combined association. (Some lichenized genera unknown to form ascocarps are necessarily described only on the basis of vegetative characters, for example *Cystocoleus, Endocena, Lepraria, Leproplaca, Leprocaulon, Phyllophiale, Racodium, Siphula* and *Thamnolia*, but these are exceptional.) As both the vegetative structures and chemical products in a lichen are known in some cases to be controlled or modified by the presence of an algal partner, it will be evident that many "lichen" genera, particularly amongst the macrolichens, are based on two of Dughi's (1954) sets of characters: (1) from the mycobiont (essentially the ascocarps), and (2) from the association. These have been referred to as "taxonomic" and "biological" respectively and once gave rise to a debate as to the theoretical acceptability of the latter.

As indicated elsewhere (Hawksworth, 1974), I consider that genera in both lichenized and non-lichenized fungi should ideally be separated by several distinct and unrelated characters. In practice this tenet has been adhered to more rigorously by lichenologists than many mycologists. Lichenologists have tended to employ often unwieldy genera comprising several hundred species subdivided into well-marked subgenera and sections rather than raise infrageneric categories to generic rank (e.g. *Bacidia, Caloplaca, Lecanora, Lecidea, Physcia, Parmelia* and *Usnea*). Since 1965 there has been a major departure from this attitude and an increasing tendency to question many long-accepted generic limits; so marked has this movement been that Poelt (1974b) reported "The concept of genus in lichens is now in a state of flux". The strongest case for generic re-modelling arises when ascocarpic features are entirely correlated with differences in vegetative structure and/or chemical products and no species showing intermediate features are known. This situation holds, for example, in the acceptance of both *Anaptychia* and *Heterodermia* (Poelt, 1965; Swinscow and Krog, 1976a), the separation of *Cetrelia* and *Platismatia* (Culberson and Culberson,

1968), and the recognition of *Bryoria* and *Sulcaria* as distinct from *Alectoria* (Brodo and Hawksworth, 1977). The correlation between pycnidial/conidial features and other characters carries equivalent weight, as in the segregation of *Phaeophyscia* from *Physcia* (Moberg, 1977). That such separations are being made increasingly now reflects the much greater attention lichens have received in recent years; this has brought a wealth of previously overlooked information to light and enabled more natural units to be recognized.

It is, however, less easy to decide if generic separations should be made on the basis of vegetative or chemical characters alone. This situation has arisen in *Parmelia* where segregates have been treated at generic rank by weighting characters of these types (e.g. Hale, 1974b). Here this occasionally seems to separate taxa similar in unweighted features and consequently many lichenologists reject this approach. A problem with weighting a vegetative character has also arisen in the distinction between *Cladina* and *Cladonia*; here the occurrence of some intermediate taxa (see Ahti, 1977) inevitably casts doubt on the desirability of the segregation.

The introduction of new or the resurrection of long-forgotten generic names for previously well-known groups of species is not something to be undertaken lightly, as the consequent disruption, particularly if the changes are not accepted by lichen systematists at large, leads to unnecessary instability in names and causes confusion not only in works of taxonomists but also to ecologists and all others who use lichen names.

Species concepts

The concept of a "species" in lichenized fungi ideally, as with vascular plants, is based on clear discontinuities in several unrelated characters (Hawksworth, 1974). In many genera of microlichens (i.e. crustose and leprose species) these characters are often specifically fungal (i.e. ascocarp derived) whilst in most macrolichens (i.e. foliose and fruticose species) the features of the vegetative thallus and chemical products assume paramount importance; furthermore, in many macrolichens ascocarps are quite unknown and in others occur only exceptionally.

In the absence of the ability to culture intact lichen thalli readily under controlled conditions, determinations of the extent of intergradations and the effects of environmental factors on thallus form have to be made on the basis of extensive field work and examination of large numbers of herbarium specimens (e.g. at least 14,000 by Degelius, 1954; 9000 by Brodo and Hawksworth, 1977; 20,000–30,000 by Runemark, 1956). As the types of environmental effects on characters employed in lichen systematics can now be considered reasonably well known (Hawksworth, 1973; Weber, 1977), the elucidation of them is more a practical rather than a fundamental problem in modern lichen taxonomy and so need not concern us further here.

The "species pair" (*Artenpaare*) hypothesis developed by Poelt (1970, 1972) suggests that extant lichenized taxa reproducing only by asexual methods ("secondary species") have arisen from otherwise identical extinct or extant species reproducing mainly or solely by sexual methods ("primary species"), the chemical components usually remaining unchanged. This hypothesis has already proved an extremely valuable aid in the understanding of speciation in lichens (e.g. Culberson and Hale, 1973; Culberson, 1973). As might be expected if a hypothesis holds true for the past, instances arise today of secondary species apparently in the process of evolution; this is seen in the occurrence of sorediate and fertile specimens of *Bryoria fremontii* (Tuck.)Brodo & D.Hawksw. (Brodo and Hawksworth, 1977) and in *Buellia subcanescens* Wern. being incompletely delimited from *B. canescens* (Dicks.)de Not. (Llimona *et al.*, 1976). Whether any taxonomic separation should be made in cases of incomplete delimitation, and at what rank, must necessarily depend on the extent to which the vicariant taxa are clearly segregated.

Chemical characters have been employed extensively in lichen systematics for over a century but, while their value at the familial and generic levels is now indisputable, there are cases at specific and infraspecific ranks where the most appropriate taxonomy remains a matter of debate. As this aspect of lichen systematics has recently been reviewed elsewhere (Hawksworth, 1976b) I do not propose to return to this here except to draw attention to some particularly complex situations which have come to light in some groups of *Usnea* species in Africa (Swinscow and Krog, 1975, 1976b).

Culberson and Hale (1973) suggested that some form of ancient hybridization of chemically different types and subsequent parallel evolution had occurred in *Parmelia* sect. *Hypotrachyna* and that this hybridization could have occurred without recombination if the pertinent genes were on homologous or additional different chromosomes. In the course of investigations into alectorioid genera in North America, Brodo and Hawksworth (1977) found three instances where rare specimens were encountered with apparently hybrid chemistries and, in two of these three cases, also intermediate morphologies. Whether such plants arise as a result of genetic recombination is necessarily a matter for conjecture as it is conceivable that: (1) propagules from different species might simply grow together sharing a single alga to form a composite plant, i.e. a "mechanical hybrid" (Henssen and Jahns, 1973); (2) two different algal species or strains might become involved with a single ·mycobiont and exert modifying effects (*see further below*); or (3) heterokaryosis might have arisen [this in turn could give rise to recombination through the "parasexual cycle", but the extent of the occurrence of this phenomenon in non-lichenized fungi is uncertain (Burnett, 1975)]. Brodo and Hawksworth (1977) did not treat their rarely occurring "hybrids" as separate taxa but in some other cases it is theoretically possible that such plants are sufficiently frequent and self-perpetuating to merit formal taxonomic recognition. Whatever the basis or bases of such hybrid-like specimens may eventually prove to be, this aspect is now something which requires consideration by lichen monographers.

A particularly neglected field of lichen systematics is the specific identification of the phycobionts (for a compilation of the meagre data on these, see Ahmadjian, 1967b) which generally requires that they are isolated in pure culture (Tschermak-Woess, 1976). The only monograph to attempt in detail to wrestle with this problem appears to be that of Degelius (1954) on *Collema*; here differences in *Nostoc* cultures occurred both between some single *Collema* species and within others. Wang-Yang and Ahmadjian (1972) cultured *Trebouxia* from different specimens of *Cladonia rangiferina* (L.)Web. and *Parmelia caperata* (L.)Ach. and showed that single lichen species could include different algal species or strains. These studies indicate that at least some mycobionts are not specific about the phycobionts with which they can form associations and that variations in the alga utilized may not affect the gross morphology of the resultant association. Conversely, it is now equally certain that in other instances the algal partner with which a mycobiont forms an association can have profound effects on the nature of the thallus formed. The strongest evidence for this phenomenon arises in cases where a single mycobiont can combine with either a blue-green or a green alga. That such a change in algal type could influence the morphology of the intact thallus has been recognized for many years in the case of blue-green alga-containing structures termed "cephalodia" arising on the thalli of lichens with a green alga as the principle phycobiont (James and Henssen, 1976). On the basis of extensive critical anatomical and chemical investigations, James and Henssen (1976) have now conclusively demonstrated not only that the same mycobiont can form morphologically quite dissimilar thalli (e.g. foliose instead of fruticose; see Fig. 3) with different types of algae, but also that in some cases these can live independently, have

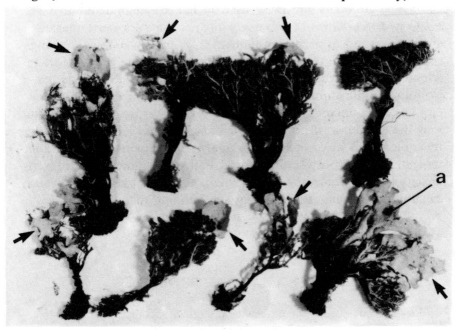

FIG. 3. *Sticta filix* (Räusch.)Nyl., an example of a lichen in which the same mycobiont produces quite dissimilar morphological structures (phycotypes) with different types of algae. Note the coralloid blue-green algal phycotype (*Dendriscocaulon* sp.) and the leaf-like lobes of the green algal phycotype (*S. filix* s.s., black arrows; *a*, apothecium restricted to the green algal phycotype) are combined in some specimens of this collection (New Zealand, South Island, Fiordland, Lake Te Anau, head of Middle Fiord above Lake Hankinson near western shore of Lake Thomson, on mossy boulder, 1962, *P. W. James*, BM). Approx life size Photograph by D. W. Fry.

sometimes been classified in different "genera" of lichens, and have different chemical products.

While the firmest evidence for algal influence on the form a lichenized association can take comes from examples where quite different groups of algae are involved (i.e. green *vs.* blue-green), as variations in algal species or strains are also known to occur within single lichen species (*see above*), it is clear that the included alga may be much more important in determining the nature of the resultant lichen than hitherto assumed. James and Henssen (1976) postulated that differences in phycobionts might perhaps be the cause of the often bewildering range of variation in both chemical and morphological features seen in some groups of *Usnea*. [The phycobiont appears necessary for the biosynthesis of at least some of the lichen products routinely used in lichen chemotaxonomy; see Hawksworth (1976b).] Such a lack of specificity for particular algal species or strains is perhaps not unexpected in view of the wide range of nutritional types to which mycobionts in some genera have become adapted. It will be appreciated that these considerations have profound implications for the systematics of lichenized fungi and re-emphasize that many characters routinely used by lichen taxonomists may be products of the dual association and not only of the mycobiont. The most satisfactory treatment of distinct morphological types produced by a single mycobiont with different algae will vary according to the extent to which they remain quite separate; nomenclatural problems surrounding such "algal phases", or "phycotypes" as they have been aptly termed by Swinscow (1977), are discussed below.

NOMENCLATURAL PROBLEMS

For nomenclatural purposes names given to lichens shall be considered as applying to their fungal components. . . . The group to which a name is assigned for the purpose of this Article is determined by the accepted taxonomic position of the type of the name. (Stafleu *et al.*, 1972: Art. 13 *p.p.*)

Starting-point date

Whereas the nomenclature of lichens is considered to start on 1 May 1753 (Linnaeus, 1753), that of most non-lichenized fungi does not commence until 1 January 1821 (Fries, 1821) according to the current *International Code of Botanical Nomenclature* (Stafleu *et al.*, 1972). Resultant from the sentences of Article 13 of this Code cited above and the enormous problems surrounding the definition of the lichenized state reviewed earlier in this contribution, are difficulties or inconsistencies in nomenclature due to these disparate starting-point dates. These arise in the following five situations:

1. Where a single genus circumscribed on the basis of the fungal partner includes consistently lichenized *and* consistently non-lichenized species. As "group" for nomenclatural purposes is determined by the type of each taxon, different starting-point dates are operative for different species within the single genus (Santesson, 1960; Hawksworth, 1975; Ahti, 1976). (Salisbury (1976) considered that "Nomenclaturally anything included in a lichen genus is a lichen, in a fungal genus is a fungus" and that "Biological definitions are irrelevant except for generic names"; but this partly circular argument is, as emphasized by Jorgensen (1977), quite unacceptable under the existing Code.)

2. In cases where it is uncertain if a taxon forms a biotrophic association with an alga, should 1753 or 1821 be adopted? There appears to be no solution to such instances under the present Code.

3. If a fungus traditionally regarded as lichenized is, as a result of detailed study, proved to be either only fortuitously associated with an alga or to be an algal parasite, the starting-point date is immediately put forward 68 years to 1821; this can result in a change in both generic name and the specific epithet (or at least in the author citation) as a result of a biological and not a taxonomic discovery. (The reverse of this situation also applies.)

4. With facultatively lichenized species should 1753 or 1821 be employed or should a different date (and consequently the possibility of a different name) be employed depending on whether or not algal cells can be demonstrated in each specimen considered?

5. For nomenclatural purposes it is not clear under the current Code whether a fungal parasymbiont of a lichen should be treated as a lichenized or as a non-lichenized fungus.

The primary object of a Code of nomenclature is to promote stability in names. In practice systematists working in the above problematical areas have followed the spirit of the Code and not tended to introduce different names or author citations depending on whether algal cells can be demonstrated in a particular specimen or not. It is also perhaps fortuitous that as many such fungi are small they were in any case not described until after 1821, although this consideration does not apply in case (1) above. The Code should cover all types of situation which can be foreseen and in this area it is clear that it does not.

To me the only solution to all the above difficulties appears to be to employ 1 May 1753 as the starting-point date for all fungi, whether lichenized or not. This need for a common starting point was asserted at the Stockholm Congress in 1950 (Ahlner, 1953), but the implications of such a fundamental change have only recently started to be assessed in detail by mycologists at large, albeit mainly for other reasons; a resumé of deliberations of a standing committee of the International Mycological Association established to consider starting-point dates in fungi has recently been prepared (Petersen, 1977). The case for 1753, provided that some safeguards proposed to protect fungal names in current later starting-point books are accepted, now seems extremely strong and it is to be hoped that a change will be effected in the Code at the next International Botanical Congress.

Interestingly, in the only comprehensive treatment of a large group comprising both lichenized and non-lichenized species to have appeared recently, Sherwood (1977) adopted 1753 for all species contrary to the present Code in the firm belief ". . . that there is no other reasonable way to deal with the nomenclature of a group which contains both lichenized and non-lichenized members". This use of 1753 by a mycologist in a largely non-lichenized group is an innovation in the starting-point debate.

Biological vs. taxonomic genera

As many lichen taxa are based on characters of the combined association and not on those of the mycobiont alone, a separate series of taxonomic names based only on the fungal components was proposed by Thomas (1939) and Ciferri and Tomaselli (1953) to replace the essentially biological names in general use; most such names were coined by adding the suffix "*-myces*" to the existing biological names (e.g. *Alectoriomyces* for the fungal component of *Alectoria*). This procedure has little to commend it (see Santesson,

1954) and is unacceptable under the current Code (Art. 13(d); pertinent sentence cited above). This dual system of names is nevertheless occasionally still used for mycobionts in pure culture (e.g. Werner, 1976), although some of its once strongest advocates no longer support it (notably Tomaselli, 1975).

To reject this approach is, however, not to deny that if one interpreted this part of the Code as meaning that only fungal characters were acceptable in the delimitation of lichenized taxa, lichen systematics as conducted today would become unacceptable; this point was made by Culberson (1961) and its significance is strengthened by the subsequent evidence for phycobial influence on thallus form and, at least in some cases, chemistry. These difficulties can be surmounted by the appreciation that a Code of *nomenclature* cannot prescribe a *taxonomy* but only aims to regulate the application of names to the units taxonomy recognizes as meriting names. In delimiting his taxa, a taxonomist is free to draw on any characters he considers pertinent regardless of their nature.

The restriction of names given to lichens for nomenclatural purposes to the mycobiont should be maintained as, by precluding the possibility of a dual system of generic names, it promotes stability in the nomenclature of lichenized associations. Further, it permits separate binomials to be employed for the algal partners of lichens.

Phycotypes

In addition to presenting taxonomic problems (see pp. 230–232), the formation of dissimilar plants by a single mycobiont depending on whether the phycobiont is a blue-green or a green alga gives rise to some nomenclatural difficulties. Dughi (1954), who had appreciated this phenomenon in the 1930s, proposed a system of quadrinomial nomenclature (e.g. *Lobaria amplissima* . *Nostoc* sp., *L. amplissima*. gonidie trebouxioïde), but this appears unnecessarily cumbersome. James and Henssen (1976) considered that in instances where only one of the phycotypes formed ascocarps it was most satisfactory to simply treat one as an algal phase of the other employing the earliest available name in species rank available (e.g. the blue-green alga-containing species previously named *Sticta dufourii* Del. would be referred to as the blue-green algal phase of *S. canariensis* Bory ex Del., which has a green phycobiont). These authors were, however, uncertain as to the treatment to recommend in two further instances: (1) where the fertile phycotype of a blue-green algal phycotype was unknown or may once have occurred but is now extinct (e.g. *Dendriscocaulon intricatulum* (Nyl.)Henss.), and (2) where both phycotypes formed ascocarps and combined phycotypes though known were very rare (e.g. some species of *Peltigera*).

In view of the wide range in phycotype situations from cases where independent plants are never produced (e.g. cephalodia in *Placopsis gelida* (Nyl.)Linds.) to pairs of phycotypes only exceptionally linked and for the bulk of their ranges acting as distinct species, it does not seem possible to formulate a single provision to include in the Code which will adequately deal with this situation. It might perhaps be thought that the inclusion of lichens in Article 59, which deals with perfect–imperfect relationships in fungi, could be applied in a modified form, but this is not recommended in view of other difficulties this could have for lichen nomenclature (*see below*).

Consequently, I feel that it is preferable to leave this situation as one for taxonomic judgement rather than nomenclatural legislation. This will enable two phycotypes to be united under one name (the earliest regardless of whether it produces ascocarps or not) or retained as distinct species as seems most expedient in each case. There would then be no *nomenclatural* objection to, for example, speaking of the *Dendriscocaulon* phase (or phycotype) of *Lobaria amplissima* (Scop.)Forss. as long as the entities were treated as *taxonomically* distinct.

Imperfect vs. perfect states

Article 59 of the current Code permits the usage of separate binomials for imperfect states (i.e. not forming ascocarps or basidia) of non-lichenized fungi in works referring to those states. At present, lichen-forming fungi are specifically excluded from this Article but it may be of interest to draw attention to implications which its use in them would have in view of it being considered as perhaps usable for phycotypes and imperfect states of lichens.

If a lichen species occurred in three states (e.g. sterile, pycnidial and ascocarpic), each of these states could then be accorded a separate name as long as when a collective name was employed it was the earliest for the sexual sporocarp-forming state. The details of the

Article are complex and currently a matter of debate amongst mycologists (Hawksworth and Sutton, 1974a, 1974b; Weresub *et al.*, 1974); most need not be cited here, but particularly pertinent is that which rules that any epithet placed in a genus typified by a specimen with the perfect state must also produce that state. As the type specimens of many macrolichens particularly lack the perfect state (even though these may be known in other collections of the same species) it will be evident that this would lead to a great deal of nomenclatural instability. Further, such instability is hard to justify as many sterile or imperfect macrolichens can be confidently assigned to particular perfect-state genera on the basis of their vegetative structures and chemistries.

In rejecting any suggestion that lichenized taxa should be covered by Article 59, it is nevertheless important to appreciate that cases more nearly akin than the situations mentioned above to those which the Article was designed to overcome do arise in the lichens. Laundon (1963), for example, took up the pycnidia-producing (imperfect state) name *Pyrenotea vermicellifera* Kunze and combined it into *Opegrapha* for the perfect state-forming taxon previously called *O. fuscella* (Fr.)Almb. In addition there are a considerable number of mainly foliicolous imperfect lichenized taxa known (see, e.g., Batista, 1961; Batista and Maia, 1965; Funk, 1973) the names of some of which may eventually be found to compete with perfect state names. In the case of the lichen-forming fungi, to have to face the prospect of such changes appears to be a small price to pay for the relative stability in the application of many other names achieved by the exclusion of lichenized taxa from the terms of reference of Article 59.

ACKNOWLEDGEMENTS

I am indebted to various colleagues for stimulating discussions in connection with various topics reviewed here over the years, particularly Mr P. W. James and Dr T. D. V. Swinscow, and also to Dr Luella K. Weresub with whom through correspondence and discussion on non-lichenological matters I appreciated the *nomenclatural* value of distinguishing taxonomy from nomenclature.

REFERENCES

Ahlner, S. (1953). Some aspects of nomenclature and taxonomy of lichens. *In* "Proceedings of the Seventh International Botanical Congress, Stockholm 1950" (H. Osvald and E. Åberg, eds), p. 809. Almqvist and Wiksell, Stockholm.

Ahmadjian, V. (1967a). "The Lichen Symbiosis." Blaisdell Publishing, Waltham, Mass.

Ahmadjian, V. (1967b). A guide to algae occurring as lichen symbionts: isolation, culture, cultural physiology, and identification. *Phycologia* **6**, 127–160.

Ahmadjian, V. (1970). The lichen symbiosis: its origin and evolution. *Evolut. Biol.* **4**, 163–184.

Ahti, T. (1976). Views. *Internat. Lichen Newsl.* **9**(1), 18–19.

Ahti, T. (1977). The *Cladonia gorgonina* group and *C. giganiea* in East Africa. *Lichenologist* **9**, 1–15.

Ainsworth, G. C. (1971). "Ainsworth & Bisby's Dictionary of the Fungi", 6th edition. Commonwealth Mycological Institute, Kew.

von Arx, J. A. and Müller, E. (1975). A re-evaluation of the bitunicate ascomycetes with keys to families and genera. *Stud. mycol., Baarn* **9**, 1–159.

Barr, M. E. (1976). Perspectives in the Ascomycotina. *Mem. N.Y. bot. Gdn.* **28**, 1–8.

de Bary, A. (1879). "Die Erscheinung der Symbiose." Strasbourg.

Batista, A. C. (1961). Um pugilo de gêneros novos de liquens imperfeitos. *Publnes Inst. mic. Recife* **320**, 1–31.

Batista, A. C. and Maia, H. da S. (1965). Alguns novos generos de liquens imperfeitos assinalados no IMUR. *Atas Inst. Mic. Recife* **2**, 351–373.

Bigelow, H. E. (1970). *Omphalina* in North America. *Mycologia* **62**, 1–32.

Bourrelly, P. (1972). "Les Algues d'Eau Douce." Vol. 1. "Les Algues Vertes." Éditions N. Boubée and Cie, Paris.

Brodo, I. M. and Hawksworth, D. L. (1977). *Alectoria* and allied genera in North America. *Op. bot. Soc. bot. Lund.* **42**, 1–164.

Burnett, J. H. (1975). "Mycogenetics." J. Wiley and Sons, London.

Cain, R. F. (1972). Evolution of the fungi. *Mycologia* **64**, 1–14.

Chadefaud, M., Letrouit-Galinou, M.-A. and Janex-Favre, M.-C. (1969). Sur l'origine phylogénétique et évolution des ascomycètes des lichens. *Bull. Soc. bot. Fr., Mém.* 1968 (Coll. Lich.), 79–111.

Church, A. H. (1921). The lichen as transmigrant. *J.Bot., Lond.* **59**, 7–13, 40–46.

Ciferri, R. and Tomaselli, R. (1953). Saggio di una sistematica micolichenologica. *Atti Ist. bot. Univ. lab. crittogam. Pavia, ser.* 5, **10**, 26–84.

Cribb, A. B. and Cribb, J. W. (1960). Marine fungi from Queensland—III. *Pap. Dep. Bot. Univ. Qd* **4**(2), 39–44.

Culberson, C. F. and Hale, M. E. (1973). Chemical and morphological evolution in *Parmelia* sect. *Hypotrachyna*: product of ancient hybridization? *Brittonia* **25**, 162–173.

Culberson, W. L. (1961). Proposed changes in the international code governing the nomenclature of lichens. *Taxon* **10**, 161–165.

Culberson, W. L. (1973). The *Parmelia perforata* group: niche characteristices of chemical races, speciation by parallel evolution, and a new taxonomy. *Bryologist* **76**, 20–29.

Culberson, W. L. and Culberson, C. F. (1968). The lichen genera *Cetrelia* and *Platismatia* (Parmeliaceae). *Contr. U.S. natn. Herb.* **34**, 449–558.

Degelius, G. (1954). The lichen genus *Collema* in Europe. *Symb. bot. upsal.* **13**(2), 1–499.

Dobbs, G. (1970). The phycotrophic fungi. *Lichenologist* **4**, 323–325.

Dughi, R. (1954). Sur la taxinomie des lichens. *Bull. Soc. Hist. nat. Toulouse* **89**, 97–120.

Duncan, U. K. (1970). "Introduction to British Lichens." T. Buncle, Arbroath.

Farrar, J. F. (1976). Ecological physiology of the lichen *Hypogymnia physodes*. II. Effects of wetting and drying cycles and the concept of "physiological buffering". *New Phytol.* **77**, 105–113.

Fries, E. M. (1821). "Systema Mycologicum." Vol. 1. Lund.

Funk, A. (1973). *Microlychnus* gen.nov., a lichenized hyphomycete from western conifers. *Can. J. Bot.* **51**, 1249–1250.

Galun, M. (1970). "The Lichens of Israel." Israel Academy of Sciences and Humanities, Jerusalem.

Hafellner, J. and Poelt, J. (1976). *Rhizocarpon schedomyces* spec.nov., eine fast delichenisierte parasitische Flechte, und seine Verwandten. *Herzogia* **4**, 5–14.

Hale, M. E. (1957). "Lichen Handbook." Smithsonian Institution, Washington DC.

Hale, M. E. (1974a) ["1973"]. Growth. *In* "The Lichens" (V. Ahmadjian and M. E. Hale, eds), pp. 473–492. Academic Press, New York and London.

Hale, M. E. (1974b). *Bulbothrix, Parmelina, Relicina*, and *Xanthoparmelia*, four new genera in the Parmeliaceae (Lichenes). *Phytologia* **28**, 479–490.

Harris, R. C. (1973). The corticolous pyrenolichens of the Great Lakes region. *Mich. Bot.* **12**, 3–68.

Hawksworth, D. L. (1972). The natural history of Slapton Ley Nature Reserve IV. Lichens. *Fld Stud.* **3**, 535–578.

Hawksworth, D. L. (1973). Ecological factors and species delimitation in the lichens. *In* "Taxonomy and Ecology" (V. H. Heywood, ed.), pp. 31–69. Academic Press, London and New York.

Hawksworth, D. L. (1974). "Mycologist's Handbook." Commonwealth Mycological Institute, Kew.

Hawksworth, D. L. (1975). Notes on British lichenicolous fungi, I. *Kew Bull.* **30**, 183–203.

Hawksworth, D. L. (1976a). New and interesting microfungi from Slapton, South Devonshire: Deuteromycotina III. *Trans. Br. mycol. Soc.* **67**, 51–59.

Hawksworth, D. L. (1976b). Lichen chemotaxonomy. *In* "Lichenology: Progress and Problems" (D. H. Brown, D. L. Hawksworth and R. H. Bailey, eds), pp. 139–184. Academic Press, London and New York.

Hawksworth, D. L. (1977a). Taxonomic and biological observations on the genus *Lichenoconium* (Sphaeropsidales). *Persoonia* **9**, 159–198.

Hawksworth, D. L. (1977b). Three new genera of lichenicolous fungi. *Bot. J. Linn. Soc.* **75**, 195–209.

Hawksworth, D. L. and Sutton, B. C. (1974a). Article 59 and names of perfect state fungi in imperfect state genera. *Taxon* **23**, 563–568.

Hawksworth, D. L. and Sutton, B. C. (1974b). Comments on Weresub, Malloch and Pirozynski's proposal for Article 59. *Taxon* **23**, 659–661.

Henssen, A. and Jahns, H. M. (1973) ["1974"]. "Lichenes. Eine Einführung in die Flechtenkunde." G. Thieme, Stuttgart.

Henssen, A. and Kowallik, K. (1976). A note on the mycobiont of *Coriscium viride* (Ach.)Vain. *Lichenologist* **8**, 197.

Hertel, H. (1969). *Arthonia intexta* Almq., eine vielfach verkannter fruchtkorperlöser Flechtenparasit. *Ber. dtsch. bot. Ges.* **82**, 209–220.

Hertel, H. (1970). Parasitische lichenisierte Arten der Sammelgattung *Lecidea* in Europa. *Herzogia* **1**, 405–438.

James, P. W. (1965). A new check-list of British lichens. *Lichenologist* **3**, 95–153.

James, P. W. and Henssen, A. (1976). The morphological and taxonomic significance of cephalodia. *In* "Lichenology: Progress and Problems" (D. H. Brown, D. L. Hawksworth and R. H. Bailey, eds), pp. 27–77. Academic Press, London and New York.

Janex-Favre, M.-C. (1971). Recherches sur l'ontogénie, l'organisation et les asques de quelques pyrénolichens. *Revue bryol. lichén.* **37**, 421–650.

Jones, E. B. G. (1976). Lignicolous and algicolous fungi. *In* "Recent Advances in Aquatic Mycology" (E. B. G. Jones, ed.), pp. 1–49. Elek Science, London.

Jørgensen, P. M. (1977). More on nomenclatural starting-points. *Internat. Lichen Newsl.* **10**(1), 13.

Keissler, K. von (1930). Die Flechtenparasiten. *Rabenh. Krypt.-Fl.* **8**, i–ix, 1–712.

Kohlmeyer, J. (1967). Intertidal and phycophilous fungi from Tenerife (Canary Islands). *Trans. Br. mycol. Soc.* **50**, 137–147.

Kohlmeyer, J. (1974). Higher fungi as parasites and symbionts of algae. *Veröff. Inst. Meeresforsch. Bremerh., Suppl.* **5**, 339–356.

Kohlmeyer, J. (1975). New clues to the possible origin of ascomycetes. *BioScience* **25**, 85–93.

Kohlmeyer, J. and Kohlmeyer, E. (1972). Is *Ascophyllum nodosum* lichenized? *Botanica Mar.* **15**, 109–112.

Korf, R. P. (1973). Discomycetes and Tuberales. *In* "The Fungi" (G. C. Ainsworth, F. K. Sparrow and A. S. Sussman, eds), Vol. IVA, pp. 249–319. Academic Press, New York and London.

Kreisel, H. (1969). "Grundzüge eines natürlichen Systems der Pilze." G. Fischer, Jena.

Laundon, J. R. (1963). The taxonomy of sterile crustaceous lichens in the British Isles. 2. Corticolous and lignicolous species. *Lichenologist* **2**, 101–151.

Letrouit-Galinou, M.-A. (1968). The apothecia of the discolichens. *Bryologist* **71**, 297–327.

Letrouit-Galinou, M.-A. (1973). Les asques des lichens et le type archeasé. *Bryologist* **76**, 30–47.

Linnaeus, C. (1753). "Species Plantarum", Vol. 2. Stockholm.

Llimona, X., Werner, R. G., Lallemant, R. and Boissiere, J. C. (1976). A propos du *Buellia subcanescens* R. G. Werner espèce primaire due *Buellia canescens* (Dicks.)D.N. *Revue bryol. lichén.* **42**, 617–635.

Luttrell, A. S. (1951). Taxonomy of the pyrenomycetes. *Univ. Mo. Stud.* **24**(3), 1–120.

Luttrell, A. S. (1973). Loculoascomycetes. *In* "The Fungi" (G. C. Ainsworth, F. K. Sparrow and A. S. Sussman, eds), Vol. IVA, pp. 135–219. Academic Press, New York and London.

Moberg, R. (1977). The lichen genus *Physcia* and allied genera in Fennoscandia. *Symb. bot. upsal.* **22**(1), i–vii, 1–108.

Moser-Rohrhofer, M. (1975). "Physiologische und vergleichende Anatomie der Flechtenpilze", Vol. 1 (Tafelband). Akademische Druck- und Verlagsanstalt, Graz.

Nannfeldt, J. A. (1932). Studien über die Morphologie und Systematik der nichtlichenisierten inoperculaten Discomyceten. *Nova Acta R. Soc. Scient. upsal.*, *ser.* 4, **8**(2), 1–368.

Oberwinkler, F. (1970). Die Gattungen der Basidiolichenen. *Votr. GesGeb. Bot.[Dtsch. bot. Ges.]*, N.F. **4**, 139–169.

Orton, P. D. (1977). Notes on British agarics: V. *Kew Bull.* **31**, 709–721.

Ozenda, P. (1963). Lichens. *In* "Handbuch der Pflanzenanatomie" (W. Zimmermann and P. Ozenda, eds), Vol. 6(9), pp. i–x, 1–199. Gebrüder Borntraeger, Berlin-Nikolassee.

Palm, B. T. (1932). Clavarien und Algen. *Svensk bot. Tidskr.* **26**, 175–190.

Petersen, R. H. (1967). Notes on clavarioid fungi. VII. Redefinition of the *Clavaria vernalis-C. mucida* complex. *Am. Midl. Nat.* **77**, 205–221.

Petersen, R. H. (1977). Starting points for nomenclature of fungi: a primer for IMC2. *Taxon* **26**, 310–321.

Pirozynski, K. A. and Malloch, D. W. (1975). The origin of land plants: a matter of mycotrophism? *BioSystems* **6**, 153–164.

Poelt, J. (1965). Zur Systematik der Flechtenfamilie Physciaceae. *Nova Hedwigia* **9**, 21–32.

Poelt, J. (1970). Das Konzept der Artenpaare bei den Flechten. *Votr. GesGeb. Bot.[Dtsch. bot. Ges.]*, N.F. **4**, 187–198.

242   D. L. HAWKSWORTH

Poelt, J. (1972). Die taxonomische Behandlung von Artenpaare bei den Flechten. *Bot. Notiser* **125**, 77–81.

Poelt, J. (1974a). Die parasitische Flechte *Lecidea insidiosa* und ihre Biologie. *Pl. Syst. Ecol.* **123**, 25–34.

Poelt, J. (1974b) ["1973"]. Classification. *In* "The Lichens" (V. Ahmadjian and M. E. Hale, eds), pp. 599–632. Academic Press, New York and London.

Poelt, J. and Hafellner, J. (1976). Die Flechte *Neonorrlinia* und die Familie Arthrorhaphidaceae. *Phyton, Horn* **17**, 213–220.

Poelt, J. and Steiner, M. (1972) ["1971"]. Über einige parasitische gelbe Arten der Flechtengattung *Acarospora* (Lecanorales. Acarosporaceae). *Annl Naturhist. Mus. Wien* **75**, 163–172.

Richardson, D. H. S. (1974) ["1973"]. Photosynthesis and carbohydrate movement. *In* "The Lichens" (V. Ahmadjian and M. E. Hale, eds), pp. 249–288. Academic Press, New York and London.

Riedl, H. (1976). Die Flechte *Bacidia chlorococca* (Stenh.)Lettau und ihre Beziehungen zu Formgattungen der Fungi imperfecti. *Phyton, Horn* **17**, 337–347.

Runemark, H. (1956). Studies in *Rhizocarpon* I. Taxonomy of the yellow species in Europe. *Op. bot. Soc. bot. Lund.* **2**(1), 1–152.

Salisbury, G. (1976). Nomenclatural starting-point for non-lichenized "lichens". *Internat. Lichen Newsl.* **9**(2), 13.

Santesson, R. (1952). Foliicolous lichens I. A revision of the taxonomy of the obligately foliicolous, lichenized fungi. *Symb. bot. upsal.* **12**(1), 1–590.

Santesson, R. (1953). The new systematics of lichenized fungi. *In* "Proceedings of the Seventh International Botanical Congress, Stockholm 1950" (H. Osvald and E. Åberg, eds), pp. 809–810. Almqvist and Wiksell, Stockholm.

Santesson, R. (1954). Fungal symbionts of lichens. *Taxon* **3**, 147–148.

Santesson, R. (1960). Lichenicolous fungi from northern Spain. *Svensk bot. Tidskr.* **54**, 501–522.

Santesson, R. (1967). On taxonomical and biological relations between lichens and non-lichenized fungi. *Bot. Notiser* **120**, 497–498.

Sherwood, M. T. (1977). The Ostropalean fungi. *Mycotaxon* **5**, 1–277.

Sivanesan, A. (1972) ["1971"]. The genus *Herpotrichia* Fuckel. *Mycol. Pap.* **127**, 1–37.

Smith, A. L. and Ramsbottom, J. (1915). Is *Pelvetia canaliculata* a lichen? *New Phytol.* **14**, 295–298.

Smith, D. C. (1976). A comparison between the lichen symbiosis and other symbioses. *In* "Lichenology: Progress and Problems" (D. H. Brown, D. L. Hawksworth and R. H. Bailey, eds), pp. 497–513. Academic Press, London and New York.

Stafleu, F. A. *et al.*, eds. (1972). International Code of Botanical Nomenclature adopted by the Eleventh International Botanical Congress, Seattle, August 1969. *Regnum Vegetabile* **81**, 1–426.

Starr, M. P. (1975). A generalized scheme for classifying organismic associations. *Symp. Soc. exp. Biol.* **29**, 1–20.

Swinscow, T. D. V. (1977). Book review [Lichenology: Progress and Problems]. *Lichenologist* **9**, 89–91.

Swinscow, T. D. V. and Krog, H. (1975). The *Usnea undulata* aggregate in East Africa. *Lichenologist* **7**, 121–138.

Swinscow, T. D. V. and Krog, H. (1976a). The genera *Anaptychia* and *Heterodermia* in East Africa. *Lichenologist* **8**, 103–138.

Swinscow, T. D. V. and Krog, H. (1976b). The *Usnea articulata* aggregate in East Africa. *Norw. J. Bot.* **23**, 261–268.

Thomas, E. A. (1939). Ueber die Biologie der Flechtenbildner. *Beitr. KryptogFlora Schweiz* **9**(1), 1–208.

Tomaselli, R. (1975). The systematic position of lichens. *Archo bot. biogeogr. ital.* **51**, 3–10.

Tschermak-Woess, E. (1976). Algal taxonomy and the taxonomy of lichens: the phycobiont of *Verrucaria adriatica*. *In* "Lichenology: Progress and Problems" (D. H. Brown, D. L. Hawksworth and R. H. Bailey, eds), pp. 79–88. Academic Press, London and New York.

Vainio, E. A. (1922). Lichenographia fennica II. Baeomyceae et Lecideales. *Acta Soc. Fauna Flora fenn.* **53**(1), 1–340.

Wang-Yang, J.-R. and Ahmadjian, V. (1972). A morphological study of the algal symbionts of *Cladonia rangiferina* (L.)Web. and *Parmelia caperata* (L.)Ach. *Taiwania* **17**, 170–181.

Weber, W. A. (1977). Environmental modification and lichen taxonomy. *In* "Lichen Ecology" (M. R. D. Seaward, ed.), pp. 9–29. Academic Press, London and New York.

Weresub, L. K., Malloch, D. and Pirozynski, K. A. (1974). Response to Hawksworth and Sutton's proposals for Art. 59. *Taxon* **23**, 569–578.

Werner, R. G. (1972). Morphologie et spécificité des symbiontes lichéniques. *C. r. Congr. Socs. sav. Paris, Sect. Sci.* **93**(3), 411–415.

Werner, R. G. (1976). Champignons de lichens incrustants en culture. *Bull. trimest. Soc. mycol. Fr.* **92**, 33–56.

Wright, C. H. (1890). British hymenolichen. *J. R. microsc. Soc.* **1890**, 647.

Zahlbruckner, A. (1903–1907). Lichenes (Flechten) B. Specieller Teil. *In* "Die natürlichen Pflanzenfamilien" (A. Engler and K. Prantl, eds), Vol. 1(1*), pp. 49–249. W. Engelmann, Leipzig.

Zopf, W. (1897). Über nebensymbiose (Parasymbiose). *Ber. dtsch. bot. Ges.* **15**, 90–92.

Chapter 12

Endemic Taxa and the Taxonomist

I. B. K. RICHARDSON

INTRODUCTION

It is no coincidence that most of the "botanically interesting" areas of the world are relatively rich in endemics; or that most of the larger "taxonomically difficult" genera contain a considerable proportion of narrow endemics. Endemic taxa have a fascination for the taxonomist. They are what makes his area floristically unique; they may be very rare, often endangered. But what do they give him in return for his attentions? This is the subject of this chapter, and we will see that, while often difficult to deal with taxonomically, an appreciation of their characteristics can be of assistance to the taxonomist.

For the purpose of this chapter the discussion is mainly limited to so-called "local endemics" having a distribution restricted to, say, a particular mountain or small group of mountains, island or group of islands or some correspondingly confined area. In addition, endemism at the species level is the main object of this chapter; but much of the discussion is appropriate to endemics of a wider distribution and to those of other taxonomic ranks.

A north-temperate botanist, knowing only his own flora, may think endemism an unusual phenomenon. But many other parts of the globe tell a different story. Many small, geographically defined areas such as islands have a high proportion of endemics (all their endemics are perforce narrow): in their indigenous floras the Mascarene Islands probably have over 60% narrow endemics, the Canary Islands nearly 50%, Corsica about 35%. For comparison, two larger areas, both parts of continental blocks and therefore less effectively isolated, can be considered: they are the Cape region of South Africa and the Iberian peninsula of Europe. In both about half of their indigenous species are endemic, of which one third (over 15% of their flora, or very approximately 1000 species) could be interpreted as narrow endemics; statistics for Cape endemism are mainly extrapolated from Weimarck (1941). Within both these areas there are numerous, relatively isolated regions—usually mountainous areas—the floras of which have a much higher proportion of narrow endemics; the Sierra Nevada region of Southern Spain has approximately 36% endemism, comparable with that given above for Corsica (Favarger, 1972). Thus, the taxonomist has to contend with a considerable number of endemic taxa if he is involved with plants of these and many other regions.

But many of the statistics on endemism are exaggerated. This is due to several factors, of which there are two main aspects. The first is concerned with the thoroughness of botanical exploration. Can we be sure that an "endemic" species is not growing in one or more other, unknown localities? Today we can be fairly confident that many narrow endemics are adequately mapped, particularly in temperate regions and on many islands. But this was certainly not the case when much formal taxonomy was formulated, or when many of the great nineteenth century Floras were written. And it is not the case today in, for example, tropical forests or even such relatively accessible areas as many in the countries bordering the Mediterranean. Van Steenis (1949), after commenting on the impossibility of adequately exploring the *Flora Malesiana* region, defined a local endemic species as "one which has hitherto been found only in one single spot or island", saying that "It is therefore scientifically inadmissible to discuss tropical local-endemic species." In North Africa and Turkey new records extending the range of erstwhile "narrow endemics" are still frequently being produced.

The second aspect is concerned with our knowledge of the taxonomy of groups over wide areas. Can we be sure that two or more so-called "narrow endemics" are not the same species with a wide or discontinuous distribution? Here we must rely on monographs and revisions of genera preferably on a world scale, or at least covering large areas corresponding to the relevant floristic region. And these are unfortunately too few in number. The new regional Floras, *Flora*

Europaea, Flora Malesiana, contain many revisions on a suitable scale. Van Balgooy (1971) has quoted instances of major reductions for the Philippines after revisions of genera for or in connection with *Flora Malesiana.* For example, *Canarium* (Burseraceae) was previously said to be represented by 45 species, all of which were endemic to this region. Today the genus is considered to contain 9 species, 4 of which are endemic. Similarly, for *Capparis* (Capparaceae) the figures were 17 species (9 endemic) but are now 13 species (1 endemic); and *Ficus* (Moraceae) 144 species (116 endemic) now reduced to 87 species (20 endemic). These changes reflect improved distributional data, but most of all, broader-based taxonomic treatments.

Numbers of endemic species can be inflated for other reasons. One is parochialism by the local botanist writing a Flora of his area; he is tempted to give undue weight to differences between his representatives of a taxon and those growing elsewhere, sometimes giving them formal recognition even at the species level. Such biased treatments are often not detected until a broader survey is undertaken; many Spanish "endemics" have been sunk into more widely distributed European species in *Flora Europaea.* Many endemic species in *Flora URSS* have similarly been sunk this time because of the relatively narrow species-concept applied by the Russian taxonomists. Often a lack of facilities and opportunities for effective comparative studies led many earlier collectors to "discover" a species many times in different parts of the globe: *Flora Zambesiaca* has shown up many such false endemics. For example *Maytenus senegalensis* (Law.) Exell (Celastraceae) has 11 constituent "species" described from Europe, India and many parts of Africa: *Ipomoea rubens* Choisy (Convolvulaceae) comprises nine "species" formerly recognized from India, Java, S. Tomé, Madagascar and four African countries. And such proliferation of species was not confined to the earlier explorers of the nineteenth century; a few years ago a new endemic species of *Lathyrus* was described from Cyprus, but it was none other than the garden "sweet pea" growing as an escape.

It is not always the case that recent work results in a decrease in numbers of endemics. A study of poorly known floras such as that of the Mascarene Islands produces many new endemics. For example M. J. E. Coode (unpublished) recognizes seven endemic species of

Myrsinaceae for Mauritius, where one was previously known (Baker, 1877); and I. B. K. Richardson (unpublished) recognizes 15 endemic species of *Diospyros* (Ebenaceae) for the Mascarenes, where previously only eight were listed.

THEORETICAL CONSIDERATIONS

Endemics have long been classified into various types; Stebbins and Major (1965) and Prentice (1976) have reviewed the history of this classification and discussed relevant terminology. We are concerned here with those aspects of endemism which directly affect the taxonomist, and Fig. 1 is an attempt to show how the endemic status of a species relates to its evolutionary history.

Clearly the area covered by a species will change with time; all species start as neoendemics and finish as palaeoendemics. Between these extremes many species will cover large areas and no longer be endemics in the sense used in this discussion; but many species, even when covering their maximum area, may still be narrowly restricted. This will be by virtue of physical or physiological (ecological) barriers, and hence the term "Holoendemics" is used for such species in the figure. Examples are island or montane taxa or those with a

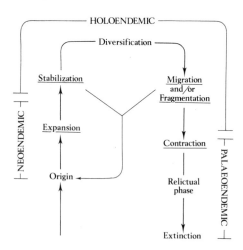

FIG. 1. The evolution of taxa in a monophyletic assemblage, showing changes in geographical area (underlined) and endemic status.

particularly specialized physiological requirement or narrow tolerance range.

So, theoretically, a knowledge of the endemic status of the species can be of great value to the taxonomist, particularly in determining the interrelations of the taxa in the group he is studying. But there are severe limitations. It will be seen that in Fig. 1 there is no time scale: taxa pass along the evolutionary pathway at different rates, and a holoendemic could be only one step removed from the ancestor of the group (i.e. have few "advanced" or "derived" characters) while a contemporaneous relic could be many steps removed (with numerous such characters). Thus, there are two major limitations to the use of endemic status as a guide in taxonomy, namely:

1. The narrow endemism of a species seen in isolation indicates nothing of its position along the evolutionary pathway. It can be a neo- holo- or palaeo-endemic.

2. Relative endemic status is no guide to phylogeny.

Despite these limitations, there are certain attributes of the three main kinds of endemic which can assist in the recognition of the status of some endemics. These are shown in Table I.

There are many exceptions to the attribute states shown, and it can be seen that there is no clear distinction between the three types. However, a combination of attributes can suggest the most likely endemic situation for a given species.

TABLE I. *The commonly occurring attribute states of the three main kinds of endemics.*

Attribute	Endemic status		
	Neo-	Holo-	Palaeo-
Taxonomically isolated	−	−	+
Geographically isolated	−	+	+
Polymorphic	−	+	−
Derived characters	+	+	−
Environment stable	±	−	+
Ploidy level high	+/−	+/−	(+)/−
Potential to expand area	+	+/−	−
Age (recent +, old −)	+	+/−	(+)/−

Further reservations regarding the equivalence of endemics stem from differences in their origin. Just as Löve (1955) pointed out the dangers of drawing parallel conclusions from both "true" and "false" vicariads (that is, essentially, species-pairs arising allopatrically or sympatrically), so the holoendemic which is confined to its area for physiological/ecological reasons is not in the same category as one which has restricted distribution for historical reasons only, and for which other suitable localities exist which it does not occupy. The former could be called "intrinsic" endemics, and their future—and perhaps considerable past—is linked to a particularly unusual combination of external factors, principally climatic, so that they may well tend to remain isolated from related stock. The latter "extrinsic" endemics have some possibility of migration from or fragmentation of their area without recourse to climatic change, as well as the relatively strong possibility of contact with a close vicariant species in a similar niche. Their potential for speciation is differently orientated and while, as Davis and Heywood (1963) put it, "classification is surely difficult enough on present-day distribution of plants without having to take into account what might happen in the future", the taxonomist should be aware of these distinctions; they are part of the present-day relationships of the taxa he is studying.

Figure 1 also shows how speciation and endemism are related, but it does not explain how the percentage endemism in the flora varies from one area to another. Areas of distribution can be fragmented by climatic or geomorphological changes to produce a disjunct species or vicariant species. This is said to account for much of the diversity of the Mediterranean flora (e.g. Davis and Hedge, 1971; Bramwell and Richardson, 1973). Davis (1971) described the way in which local endemics are distributed along the border between two floristic regions in Turkey. Valentine (1970) reviewed the case for speciation at ecotonal regions, and Stebbins (1974) discussed the evidence for most speciation occurring in semi-arid ecotonal areas. Stebbins and Major (1965) mapped the distribution and frequency of endemics in California, one of the richest areas for endemism, and account for it in the diversity of habitats and the climatic history of the region.

The species-richness and attendant endemism of some apparently relatively stable areas, notably tropical rain forests, have long intrigued biologists. Some have called such areas "Museums" (see Stebbins, 1975), implying that the local taxa are relictual; others (e.g.

Ashton, 1969) have placed much importance on competition for ecological niches as promoting speciation. But Greuter (1972), discussing the relict element in the Cretan flora, has pointed out that the ecological tolerance of a relictual species can be surprisingly wide, and considers that it is change in composition of the community which provides the main stress. Tolerance was discussed in a similar context by Gillett (1962). He stated the case for "pest pressure" being the main reason for the "apparently pointless multiplicity of species"—and its concomitant high local endemism—"in all areas in which it has had time to operate." Grubb (1977) has stressed the importance of the regeneration niche in a review of species-richness of plant communities, and current ideas on population mechanics (Taylor and Taylor, 1977) may well contribute towards an understanding of the possibilities for speciation in the so-called "stable areas"; a species with a dynamic total population structure will be more likely to survive sudden stress, perhaps by production of isolated satellite populations which could then diverge to become local vicariants. Such aspects of speciation have led to very diverse areas of the globe being well endowed with local endemics.

We have seen that both relationships between species, and speciation, are reflected in endemics in their different guises and concentrations. The two main approaches in taxonomy today are, according to Davis and Heywood (1963):

1. An empirical approach, based on the observed facts of character distribution. This results in a basic classification.
2. An interpretative approach, in which this classification is improved and interpreted in evolutionary or phylogenetic terms.

Thus endemism can be seen to be directly related to the latter, interpretative approach. The next section will show how useful in practice are data on endemics in this connection, and finally we will consider the impact of endemics on the taxonomist's empirical approach.

ENDEMICS IN THE INTERPRETATIVE APPROACH

Cracraft (1975) has concisely surveyed the prevailing ideas on historical biogeography, comparing classical Darwinian theory with phylogenetic and vicariance models. The two former are concerned with "recognizing centres of origin and pathway of dispersal" and "accepting the notion that knowledge of phylogeny is of considerable importance in reconstructing biogeographic history". By contrast vicariance models build up patterns of distribution, generalized tracks, from biogeographical data, corresponding to the historical areas of whole biota; a centre of origin is not invoked, and most vicariance (and disjunctions) require earth history rather than long distance dispersal for explanation. Once generalized tracks are accepted, aberrant vicariance of a particular taxon can be questioned and the taxonomy and phylogeny of its constituent endemics reconsidered; the approach to phylogeny is *a posteriori*, and hence of value to the taxonomist.

Dahlgren and Rao (1971) studied the affinities of *Oftia*, a small genus with endemics in southern Africa and Madagascar. Their opening words are significant:

In association with families which have highly disjunctive distributions on the southern continents (Restionaceae, Proteaceae, Cunoniaceae etc.) one will also usually find Myoporaceae . . . *Oftia* is the only genus of this family indigenous to Africa".

They were implying an aberrant vicariance pattern. As a result of their studies they concluded that the genus *Oftia* is near to *Teedia* (Scrophulariaceae), another small endemic South African genus. Unusual distribution of local endemics has here pointed the way to a change in the taxonomy.

Dahlgren (1963), in part of his revision of the 240 species of *Aspalathus* (Leguminosae—Genisteae), tentatively suggested certain local endemic species to be relicts, because of the absence of morphologically close vicariant species. For close vicariants which are relatively well isolated geographically, he said:

their area seems relict indeed in spite of the fact that species of each pair have probably been evolved from a common ancestral population (however, perhaps quite long ago).

Here endemism is giving two clues to possible phylogeny.

In Europe, also in the Leguminosae–Genisteae, the generic delimitations of *Cytisus*, *Genista* and satellite taxa have long been a source of argument. The infrageneric groupings of these two genera show differences in the geographical distribution of their constituent endemic species. In *Cytisus* the distribution patterns are of several commonly occurring types. But in *Genista* this is not so evident. A likely explanation is that the morphologically defined groups of species in *Genista* are artificial, and work in other fields supports changes in the taxonomy at this level and, as a result, perhaps higher levels. Further knowledge of the distribution patterns of these vicariant endemics might lead to a reassessment of the taxonomy of the group.

These commonly occurring distribution patterns mentioned above are equivalent to the generalized tracks of Croizat (1960). They reflect earth history, principally geological and climatological changes, and the evolution that has taken place in step with these changes. Davis (1970) has pointed out that many species of Mediterranean plants are associated with *Cedrus* (Pinaceae), a small genus with several vicariant species. Ferguson (1967), on the basis mainly of palynological studies, mapped the past distribution of *Cedrus*. During the Eocene the genus was present in Iceland and central Europe and since that time there has been a general migration southwards, producing the present disjunct distribution pattern and the vicariant species within it. Davis and Hedge (1971) discussed some of the genera which are today associated with *Cedrus*—that is, have a similar disjunct distribution of vicariants. An explanation of this is that these other genera have been sympatric with *Cedrus* for part of their history, and, once associated, have been subjected to the same vicissitudes. This pattern is a generalized track, and the taxonomy of many genera should fit in with it. This clue to taxonomy was used, for example, by Richardson (1975) in revising the Mediterranean genus *Centranthus* (Valerianaceae). Another case of association with a conifer was discussed by Stebbins and Major (1965): in the Sierra Nevada of California there is a high concentration of relatively old species in the area of distribution of *Sequoiadendron giganteum* (Lindley) Bucholz.

The endemic genera of the "Tepui" summits or "Lost worlds" of north-eastern South America and their vicariant taxa have recently been analysed (Maguire, 1971). After the Planalto of Brazil, the next closest affinities of this flora lie with tropical and Southern Africa, and only afterwards with the neighbouring Andes and West Indies. Such an analysis, and the subsequent inferences as to taxonomy and history of the taxa, is only possible because of the restricted distributions and isolation of the taxa. Such an area where three-quarters of the endemics (3000 species) of the region are confined to the mountain summits is conducive to this type of study.

Tryon (1971) analysed the distribution of fern floras of oceanic islands, and one of his findings was that:

species in the source area that are of limited distribution migrate to islands less often than those of broad distribution, but when they do they are more likely to evolve into insular endemics.

This is perhaps a special case of the founder principle. A widespread source species will repeatedly migrate to the new area and will tend to "buffer" changes in the pioneer population; a very rare introduction from a narrow endemic source will have time to diverge under weak selection pressure without such "buffering". Thus many endemics have quite likely arisen from other endemics. This could provide an alternative explanation for cases of "vicariant evolution", such as that in *Centaurea* sect. *cheirolophus* (Compositae—Cardueae) in the Canary Islands discussed by Bramwell (1972). His suggested explanation is that "species formation has probably taken place by fragmentation of a once widespread parent species . . . by volcanic activity". This is unlikely for the endemic species of an oceanic island archipelago which can only have arisen by stepwise or repeated migration from extant stock and subsequent divergence, implying polytypic rather than monotypic origin.

Cytological studies of endemics have proved useful to the taxonomist. Favarger and Contandriopoulos (1961) produced a classification of endemics based on the cytology of an endemic and its vicariant taxa:

1. Palaeoendemics: isolated systematically, old, with little variation, not necessarily having arisen in their area of survival.
2. Schizoendemics: produced by "gradual speciation", having a common origin and identical chromosome numbers.
3. Patroendemics: narrow diploids which have given rise to widely distributed polyploids.
4. Apoendemics: narrow polyploids which have arisen from widely distributed diploids.

Historical aspects of floras and the relationships of some of their constituent taxa have been surveyed using these criteria. Studies of the flora of Corsica by Contandriopoulos (1962) and of California by Stebbins and Major (1965) are two examples. In addition, surveys of the relative percentages of polyploidy of floras of different regions can produce comparative data which can be of use in discussions of the ages of floras because in general, the younger the flora, the higher the percentage of polyploids. Bramwell (1972) used the low percentage polyploidy (less than 25%) as one of the arguments for the relict nature of the indigenous Canary Islands flora; and recently Hedberg and Hedberg (1977) have surveyed polyploidy in the endemic Afro-montane flora and discussed the phenomenon in other montane regions.

Cytological studies and hybridization experiments are pursued by the taxonomist not only to investigate relationships but also to assist in determining the relative status of taxa; this is mentioned again in the final section of this chapter.

Two problematical groups of neoendemics have received particular attention in evolutionary studies. One is the so-called cryptoendemics, difficult to recognize as they closely resemble the parent species morphologically but show chromosomal distinctions; they are diploid like the relatively widespread parent, often more or less sympatric in part of its range, and are discussed with reference to the Californian flora by Lewis (1972). The other group is the seemingly paradoxical "active epibiotics" and their products (schizoendemics). Here an apparently relict species appears to have undergone a burst of diversification, but it is better considered not as a relict species in the sense of the palaeoendemics in Fig. 1, but as an old holoendemic which has remained unchanged until relatively recently. Probably the species has avoided stress conditions for a long period. An area where this phenomenon has been studied in several genera is in the Canary Islands (Bramwell, 1972) where conditions for the preservation of holoendemics have existed for millions of years; the humid forests have long remained effectively isolated from immigrant potential competitors, and, as Stebbins and Major (1965) concluded:

> In regions with ample moisture . . . floras are likely to be relatively stable, and most of the endemic species are ancient or at least not modern.

The evolutionary products of these "active epibiotics" are species which are often adapted to fit a particular habitat—hence the use of the term "adaptive radiation" for the formation of these endemics. Examples in the Canaries are in *Sonchus* (Compositae—Cichorieae) (Bramwell, 1972; Aldridge, 1977) and *Argyranthemum* (Compositae—Anthemideae) (Humphries, 1976).

The above examples have illustrated the way in which endemic taxa, with their distributional and other characteristics, can be used by the taxonomist in the interpretative approach in the study of the relationships of taxa.

ENDEMICS IN THE EMPIRICAL APPROACH

Two of the duties which the taxonomist has to perform in this field are directly influenced by endemism. One is the recognition of taxa— where neoendemics may be scarcely differentiated from the parent species. The other is in the choice of status which the endemics and often related widespread species warrant.

For the first of these, the recognition of discontinuous variation, when a new taxon is detected, we can return to an example near the end of the previous section, that of cryptoendemics. An extreme case of endemics which are difficult to detect is the results of "Quantum speciation" (Grant, 1971) discussed by Lewis (1972):

> . . . these local endemics arose relatively recently through chromosome reorganisation in extremely small, ecologically marginal populations of the more widespread species.

Lewis continues:

> How many endemics in California have arisen through quantum speciation is unknown because most of them probably have not yet been discovered. The derivative species are so similar morphologically to the parental species that they are usually indistinguishable from minor morphological variants of the parental species in the absence of chromosome counts that show an aneuploid difference or hybrids of low fertility in which meiosis can be studied.

Well documented examples are to be found in *Clarkia* (Onagraceae) and *Gilia* (Polemoniaceae), both being genera which are relatively easy to work with experimentally. Many parallel cases of endemic species are likely to remain undiscovered; some will be found accidentally by the perceptive taxonomist particularly in the course of field observations, or will come to light during biosystematic studies.

The choice of status with which to accord some endemics is very problematical for the taxonomist. Sometimes the status—mainly for psychological reasons—is too high. Two commonly occurring situations illustrate this. Firstly, there is a tendency for the taxonomist to place too much importance on the palaeoendemic which, as we have seen, is often morphologically and geographically isolated. This emphasis is often justified if the goal is an artificial or phenetic classification, but is dangerous (as seen on p. 249) if phylogeny is to be inferred. The other situation was alluded to in connection with parochialism in local Floras. When species are viewed in a wider context, such as in *Flora Europaea* (Tutin *et al.* 1964), many local endemics are found not to have a unique character-combination, and are treated as subspecies or are even sunk into synonymy: a glance at the index to *Flora Europaea* will reveal many such instances. However sometimes the herbarium taxonomist will fail to appreciate that such local endemics may indeed be distinct species which have arisen independently, because they are not morphologically distinct. Such cases are sometimes revealed by their unusual geographical distribution; some of the "species" of *Gagea* (Liliaceae) with a disjunct range in Europe may well fall into this category of taxa of polytopic origin.

Many of the problems associated with the formal recognition of endemic taxa and their appropriate rank were discussed as a result of a paper by Runemark (1971) at the 6th Flora Europaea Symposium. He reviewed some of the Lund-based projects on the flora of the central Aegean, an area relatively rich in endemics and illustrating different problems for the taxonomist. Several of these are briefly reviewed to indicate the range of problems and their interrelations. There is no doubt that parallel situations occur elsewhere; the choice of this small area of the Mediterranean is because of the particularly complex history of the area which has led to these patterns of endemism. The intensive studies on a particularly amenable group of plants, which have been aimed at an understanding of the flora of the area have often, because of their depth, produced as many problems as they have solved.

Strid (1970) reported on extensive investigations into the *Nigella arvensis* L. (Ranunculaceae) complex. There are basically two patterns of variation: one is discontinuous, between most of the larger islands of the Central Aegean; the other is continuous, across the Greek–Turkish mainland. What status could be conferred on these taxa? After extensive hybridization programmes, sterility barriers were drawn on the map and several local island species were recognized, while other island taxa were treated as subspecies of another species. The mainland taxa were divided somewhat arbitrarily into three subspecies of *N. arvensis* sensu stricto.

Runemark (1971) mentioned *Huetia* (Umbelliferae), in which variation occurs both between islands and between different mountains on the Greek mainland, and where there also seems to be an underlying altitudinal differentiation. Can these two aspects of variation be reconciled within the framework of formal categories?

Sometimes a species is well differentiated into several taxa on islands or mountain-tops, but on the adjacent mainland (or lowland) it shows a continuous variation which embraces that shown by the isolated populations. *Galanthus* (Amaryllidaceae) on some Aegean islands and on the Balkan mainland shows this pattern of variation.

And another situation, somewhat similar to the previous examples, was described by Heywood (1960) for Spanish populations of *Centaurea tenuifolia* Dufour (Compositae–Cardueae): two widespread taxa, traditionally treated as northern and southern subspecies, show some overlapping of characters. The southern taxon can be subdivided into a number of morphologically discrete mountain-top populations. Under normal circumstances these montane taxa would merit at least subspecific rank but Heywood points out that it would then be of a different order from the two traditional northern and southern subspecies.

In such cases as the preceding examples, the traditional taxonomic framework seems inadequate, although if all such cases could be investigated as thoroughly as the *Nigella arvensis* complex more objective reasons could be given in proposing classifications with their attendant formal ranks. The situation is easier if certain informal collective names can be used. Such terms as "species group" (*Flora Europaea*: ad hoc assemblage of closely related species, sympatric or allopatric), "superspecies" (Mayr, 1942: closely related allopatric species), "coenospecies" (Turesson, 1922: species which can exchange genes) and "ochlospecies" (White, 1962: species with a complex pattern of variation showing little correlation with geography) can also be used for the name of a complex. Thus Strid (1970) adopts the term "*Nigella arvensis* coenospecies".

Runemark (1971) also discusses the practicalities of recognizing all discrete island taxa formally. In a treatment such as *Flora Europaea* (Tutin et al., 1964→) it would be impracticable to recognize all the 35 populations of the extremely variable *Anthemis scopulorum* Rech. fil. (Compositae—Anthemideae) in the Aegean region as formal taxa, although almost all are morphologically distinct and variation within populations is very small; only in a monograph or a local Flora would this be acceptable.

We have seen some of the main ways in which endemism can enter into the taxonomist's empirical approach; it brings many problems, some of which are virtually impossible to deal with within the existing formal taxonomic framework.

CONCLUSIONS

We have seen some of the ways in which endemics impinge on the taxonomist's work. It is not the endemic *per se* which is exceptional compared with any other individual species—except perhaps that it may be undercollected or endangered. It is the interrelations of endemics with their related taxa which introduce special conditions: some of these the taxonomist can exploit, particularly in his studies of relationships; others bring problems, particularly in the way he orders his data in the formal representation of variation patterns.

Some aspects of endemism have not been discussed. One is apomixis and its problems. Another is the use of character-states of supposed palaeoendemics as clues to phylogeny, treating them as "primitive" or "unspecialized"; but we have seen that relative endemic status is no guide to phylogeny and that palaeoendemics can exhibit relatively numerous "derived" characters. And the need for the conservation of endemics has scarcely been touched upon: many endemics are endangered and it is an urgent task of the taxonomist to study and document these species before it is too late.

ACKNOWLEDGEMENTS

Ideas have frequently been put into perspective during discussions with Dr C. J. Humphries, British Museum (Natural History). Helpful criticism of parts of the manuscript has been made by Professor D. M. Moore, Reading University. And the Kew fund of knowledge has been tapped thanks to the generosity of many colleagues. To all, my gratitude.

REFERENCES

Aldridge, A. (1977). A critical appraisal of the Macaronesian *Sonchus* subgenus *Dendrosonchus* s.l. *Botanica Macaronesica* **2**, 25–57.

Ashton, P. S. (1969). Speciation among tropical forest trees: some deductions in the light of recent evidence. *Biol. J. Linn. Soc.* **1**, 155–196.

Baker, J. G. (1877). "Flora of Mauritius and the Seychelles." London.

Bramwell, D. (1972). Endemism in the flora of the Canary Islands. *In* "Taxonomy, Phytogeography and Evolution," (D. H. Valentine, ed.), pp. 141–159. Academic Press, London and New York.

Bramwell, D. and Richardson, I. B. K. (1973). Floristic connections between Macaronesia and the East Mediterranean region. *Mongr. Biol. Canar.* **4**, 118–125.

Contandriopoulos, J. (1962). Récherches sur la flore endémique de la Corse et sur ses origines. Thèse, Faculté des Sciences, Montpellier.

Cracraft, J. (1975). Historical biogeography and earth history: perspectives for a future synthesis. *Ann. Missouri Bot. Gard.* **62**, 227–250.

Croizat, L. (1960). "Principia Botanica." Croizat, Caracas.

Dahlgren, R. (1963). Studies in *Aspalathus*. Phytogeographical aspects. *Bot. Notiser* **116**, 431–472.

Dahlgren, R. and Rao, V. S. (1971). The genus *Oftia* Adans. and its systematic position. *Bot. Notiser* **124**, 451–472.

Davis, P. H. (1970). *Geranium* sect. *tuberosa*, revision and evolutionary interpretation. *Israel J. Bot.* **19**, 91–113.

Davis, P. H. (1971). Distribution patterns in Anatolia with particular reference to endemism. *In* "Plant Life of South-West Asia" (P. H. Davis, P. C. Harper and I. C. Hedge, eds), pp. 15–26. Botanical Society of Edinburgh.

Davis, P. H. and Hedge, I. C. (1971). Floristic links between N.W. Africa and S.W. Asia. *Annln. Naturh. Mus. Wien.* **75**, 43–57.

Davis, P. H. and Heywood, V. H. (1963). "Principles of Angiosperm Taxonomy." Oliver and Boyd, Edinburgh and London.

Favarger, C. (1972). Endemism in the montane floras of Europe. *In* "Taxonomy, Phytogeography and Evolution" (D. H. Valentine, ed.), pp. 191–204. Academic Press, London and New York.

Favarger, C. and Contandriopoulos, J. (1961). Essai sur l'endémism. *Ber. Schweiz. bot. Ges.* **71**, 384–406.

Ferguson, D. K. (1967). On the phytogeography of Coniferales in the European Cenozoic. *Palaeogeogr. Palaeoclim. Palaeoecol.* **3**, 73–110.

Gillet, J. B. (1962). Pest pressure, an underestimated factor in evolution. *Systematics Association Publication* No. 4, 37–46.

Grant, V. (1971). "Plant Speciation." Columbia University Press, New York and London.

Greuter, W. (1972). The relict element of the flora of Crete and its evolutionary significance. *In* "Taxonomy, Phytogeography and

Evolution" (D. H. Valentine, ed.), pp. 161–177. Academic Press, London and New York.

Grubb, P. J. (1977). The maintenance of species-richness in plant communities: the importance of the regeneration niche. *Biol. Rev.* **52**, 107–145.

Hedberg, I. and Hedberg, O. (1977). Chromosome numbers of afroalpine and afromontane angiosperms. *Bot. Notiser* **130**, 1–24.

Heywood, V. H. (1960). Problems of geographical distribution and taxonomy in the Iberian peninsula. *Feddes Repert.* **63**, 160–168.

Humphries, C. J. (1976). A revision of the Macaronesian genus *Argyranthemum* Webb ex Schultz Bip. (Compositae—Anthemideae). *Bull. Br. Mus. Nat. Hist. (Bot.)* **5**, 147–240.

Lewis, H. (1972). The origin of endemics in the California flora. *In* "Taxonomy, Phytogeography and Evolution," (D. H. Valentine, ed.), pp. 179–189. Academic Press, London and New York.

Löve, A. (1955). Biosystematic remarks on vicariism. *Acta. Soc. Fl. Fenn.* **72**, 1–14.

Maguire, B. (1971). On the flora of the Guayana highland. *In* "Adaptive Aspects of Insular Evolution" (W. L. Stern, ed.), pp. 63–78. Washington State University Press.

Mayr, E. (1942). "Systematics and the Origin of Species". Columbia University Press, New York.

Prentice, H. C. (1976). A Study in Endemism: *Silene diclinis*. *Biol. Conserv.* **10**, 15–30.

Richardson, I. B. K. (1975). A revision of the genus *Centranthus* DC. (Valerianaceae). *Bot. J. Linn. Soc.* **71**, 211–234.

Runemark, H. (1971). Investigations of the flora in the Central Aegean. *Boissiera* **19**, 169–179.

Stebbins, G. L. (1974). "Flowering Plants: Evolution above the species level." Arnold, London.

Stebbins, G. L. and Major, J. (1965). Endemism and speciation in the California flora. *Ecol. Monogr.* **35**, 1–35.

Strid, A. (1970). Studies in the Aegean flora, XVI. Biosystematics of the *Nigella arvensis* complex. *Opera Bot.* **28**, 1–169.

Taylor, L. R. and Taylor, R. A. J. (1977). Aggregation, migration and population-mechanics. *Nature, Lond.* **265**, 415–421.

Tryon, R. (1971). Development and evolution of fern floras of oceanic islands. *In* "Adaptive Aspects of Insular Evolution" (W. L. Stern, ed.), pp. 54–62. Washington State University Press.

Turesson, G. (1922). The species and variety as ecological units. *Hereditas* **3**, 100–113.

Tutin, T. G., Heywood, V. H. Burges, N. A., Valentine, D. H., Walters, S. M. and Webb, D. A., eds. (1964 →). "Flora Europaea" 5 vols (in progress). Cambridge University Press, Cambridge.

Valentine, D. H. (1970). Evolution at zones of vegetational transition. *Feddes Repert.* **81**, 33–39.

Van Balgooy, M. M. J. (1971). Plant-geography of the Pacific. *Blumea*, Suppl. 6. Leiden.

Van Steenis, C. G. G. J. (1949). "Flora Malesiana", Vol. 4, p. lvi. Noordhoff. Koff, Djakarta.

Weimark, H. (1941). Phytogeographical groups, centres and intervals within the Cape flora. *Lunds Univ. Årsskr. Avd.* 2, **37** (5), 1–143.

White, F. (1962). Geographic variation and speciation in Africa with particular reference to *Diospyros*. *Systematics Association Publication* No. 4, 71–103.

Chapter 13

British Endemics

S. M. WALTERS

INTRODUCTION

Writing *Flora Europaea* (Tutin *et al.*, 1964 →) has inevitably involved the reassessment of taxa described as restricted to a particular country or region of Europe, particularly when such taxa have been accepted as species in national Floras. Special interest centres on narrowly endemic species apparently confined to a single country or island group; not only are such species traditionally of great phytogeographical interest, but they are increasingly the concern of scientists who are involved in advising the organizations responsible for policies of nature conservation at national and international levels. A preliminary *List of Rare, Threatened and Endemic Plants of the Countries of Europe* has been published recently (Lucas and Walters, 1976), and provides a convenient starting point for a reassessment of the endemic element in the flora of the British Isles (including Ireland).

PREVIOUS WORK ON ENDEMISM IN THE BRITISH FLORA

The first important contribution to this subject seems to have been made by Alfred Russell Wallace nearly a century ago, who enlisted the help of H. C. Watson to provide a comment on British endemic vascular plants in his remarkable book *Island Life* (Wallace, 1895), which surveys in considerable detail the peculiarities of island faunas and floras and assesses their evolutionary significance. Wallace quotes Watson:

> It may be stated pretty confidently that there is no "species" (generally accepted among botanists as a good species) peculiar to the British Isles. True, during the past hundred years nominally new species have been named and described on British specimens only, from time to time. But these have gradually come to be identified with species described elsewhere under other names—or they have been reduced in rank by succeeding botanists, and placed or replaced as varieties of more widely distributed species.

One senses, reading the whole chapter, that Wallace was surprised, and even a little disappointed, at Watson's rather negative assessment, and it is interesting to find that, in the second edition of his book, he was able to include a list, supplied by Bennett, of no fewer then 72 British endemic taxa (species, subspecies and varieties)! Be that as it may, we have to admit that most taxonomists would see the position today as not very different from that expressed by Watson—we have very few endemic species of vascular plants which are so clearly different from their closest relatives elsewhere in the world that their status as species has not been questioned and is not likely to be.

Perhaps it is because there seemed so little to say on British endemic plants that references in the literature are so sparse. I have been unable to find any substantial discussion of the subject between Wallace and the famous paper by Wilmott (1930) on the history of the British flora just half a century later. In the section on "endemic species" in his paper, Wilmott claims there are several British endemics, saying: "it is immaterial whether they are 'big' species or 'small' species: they are plants not known to occur outside the British Isles." He dismisses the claim of several taxa to be endemic (including, interestingly enough, *Primula scotica* Hooker—*see below*), but then discusses individually "several good cases", beginning with *Fumaria occidentalis* Pugsley. The use Wilmott makes of the evidence from these endemics to argue "per-glacial survival" of elements of the British flora *in situ* need not concern us here, but his objective attempt to assess the taxonomic validity of the British endemics was very useful, and long overdue. Two years later, Salisbury (1932) considered "the endemic component" in his remarkable paper on *The East Anglian Flora*; his list of British endemic species shows an interesting lack of correspondence with those accepted by Wilmott, and emphasizes how little coordinated consideration seems to have been given to the problem. Other writers who have mentioned British endemics include Good (1974), Matthews (1955) and Turrill (1948), but none of these authors devotes more than a few paragraphs to them, or even attempts to give a complete list. Most recently, Gilbert (1967) has published a very readable popular account of *Britain's endemic flowering plants*, based upon the assessment of endemism in the *Flora of the British Isles* by Clapham *et al.* (1962); this short article is recommended, not only for its intrinsic interest as being, to my knowledge, the *only* paper wholly devoted to this subject, but also for its historical information on the recognition of particular British endemics.

THE ASSESSMENT OF ENDEMIC STATUS

It is not difficult to see why so few British botanists have written on the British endemics. There are considerable difficulties in deciding endemic status because of difficulties of taxonomic assessment, and there is the further problem of deciding whether to include the many endemic microspecies described in the few large "critical" apomictic genera, of which the most important are *Hieracium* and *Taraxacum* in the Compositae, and *Rubus* in the Rosaceae.

The problem of taxonomic assessment can be illustrated by reference to three British plants. The first is the "Isle of Man Cabbage", now known at *Rhynchosinapis monensis* (L.) Dandy ex Clapham. To Ray (1670), this plant was *"Eruca monensis laciniata lutea*: Jagged Yellow Rocket of the Isle of Man", and it was on the basis of Ray's description that Linnaeus (1753) included it in his *Species Plantarum* as *Sisymbrium monense* L. Neither Ray nor Linnaeus knew the exact extent of the species in Britain, nor would they have been surprised to learn that it occurred on the European continent. Now we can be reasonably sure that *Rhynchosinapis monensis* from coastal localities in the west of Britain does differ in the specific characters given in our Floras from all other European Cruciferae, and that no plants referable to this species can be found, for example, along the Atlantic coast of France. *Flora Europaea* (Tutin *et al.*, 1964 →) in fact distinguishes ten species of

Rhynchosinapis, of which two are British endemics—*R. monensis*, occurring from W. Scotland to Glamorgan, and the very restricted endemic *R. wrightii* (O. E. Schulz) Dandy ex Clapham, found only on Lundy Island and not described as distinct until 1936. Four of the remaining eight European species of this genus are narrow endemics restricted to a single mountain range or coastal region of W. or S. Europe.

The second British species is the widespread but local *Genista anglica* L. As in the case of the Isle of Man Cabbage, the scientific name shows its distribution as known to Linnaeus: in fact, if we look in the *Species Plantarum* (p. 710) we find the quite precise statement "Habitat in Angliae ericetis humidiusculis", and Linnaeus presumably knew it only from England. Two centuries later, we have much more precise information about its total distribution, which we now know is W. European, extending from Portugal, Spain and S.W. Italy to S. Sweden (though, curiously, omitting Ireland). The taxonomic distinction of both these British plants has stood the test of time, but in one case we know that the species is far from being restricted to Britain.

The third British plant we can use as illustration was, like both the others, quite well known to botanists before Linnaeus, and accepted by Linnaeus as a species; it is the handsome "Western Spiked Speedwell", which he called *Veronica hybrida* L., and which *Flora Europaea* now includes in the widespread and variable species *Veronica spicata* L. There seems little doubt that Linnaeus was influenced, in deciding to give it specific rank, by the fact that this taxon was very well described by Ray and others as a British plant, but his comment on distribution "Hab. in Europa rarius" suggests that he had seen similar large, robust plants from elsewhere in Europe. Here we have the opposite case from that of *Genista anglica*: a plant originally described from Britain, not thought by Linnaeus to be endemic to Britain, and demoted in modern Floras to mere varietal status.

The generalization we can make is that nearly all British "species" known to early botanists, whether originally thought to be endemic or not, have proved to be best treated as conspecific with some continental species. In fact *Rhynchosinapis monensis* is unique in being a pre-Linnaean species still accepted as an endemic species in *Flora Europaea*. The complete list of accepted (i.e. numbered) species in *Flora Europaea* which are endemic to Britain and Ireland (the geographical "British Isles") is given in the Appendix, and amounts to 23 species.

To this small but by no means negligible group we might wish to add the endemic microspecies of *Rubus*, *Sorbus*, *Hieracium* and *Taraxacum*, all genera in which we know that many local taxa distinguishable morphologically from each other are characterized by possessing apomictic (non-sexual) modes of reproduction. The difficulty is that there are so many endemic apomictic microspecies that their addition to the 23 non-critical endemics already defined would greatly distort any statistical comparison which might depend upon these figures. Such distortion is the more serious because the degree of taxonomic knowledge of critical genera such as *Hieracium* is very different in different European countries, so that statistical comparisons might be quite invalid. For this reason it is probably best to limit our main consideration to the non-critical endemic component, and consider the apomictic taxa only collectively.

THE SIGNIFICANCE OF ENDEMISM IN THE BRITISH FLORA

If we follow Good (1974) and adopt the figure of 1750 as the approximate number of native species in the British flora (excluding apomictic microspecies), we see that the percentage of endemism is only a little more than 1%. This compares with, for example, the flora of the Balearic Isles where the endemism has been calculated as about 3%, and contrasts very strongly indeed with oceanic island groups such as the Hawaiian Isles, for which a figure of *c.* 90% has been given! This general correlation between the geological age of an island group and the degree of endemism in its flora is, of course, well known, and, in view of the known Quaternary history of the British flora and vegetation, any significant endemic element in our present-day flora would indeed be difficult to explain. The British flora is essentially re-immigrant and geologically extremely recent; it therefore contains no clear cases of "palaeoendemic" species which, because of their taxonomic status, might reasonably be thought to be relict from an ancient, pre-glacial flora. Examples of palaeoendemic taxa (genera or species) are very familiar in southern Europe: one could cite *Hypericum balearicum* L. in the Balearic Isles, and the endemic Gesneriaceae in the mountains of the Pyrenees and the Balkans.

No doubt the absence of "good" old endemic species comparable to this Tertiary relict element in southern Europe explains the virtual absence of any evidence of the subfossil occurrence in Britain of endemic plants. None of the 23 listed endemics can be unambiguously identified from pollen or even macro-fossil material, and it is significant that Godwin does not even mention endemism in his standard work on the *History of the British Flora* (Godwin, 1975). Only one of the British endemics, *Primula scotica*, is in fact recorded subfossil in Britain, and that is a single, early record on which little reliance can be placed (H. Godwin, personal communication).

What then *can* we say about the British endemics? Interesting generalizations emerge when they are classified according to their known (or presumed) reproductive peculiarities. One group, represented by *Alchemilla minima* Walters [an assessment of the significance of this endemic apomict has been made by me previously (Walters, 1972)], *Calamagrostis scotica* (Druce) Druce and the *Limonium* species, belong to genera in which apomixis is proved or strongly suspected; such taxa are not essentially different from the many apomictic microspecies in *Rubus*, *Sorbus*, *Hieracium* and *Taraxacum*, and their inclusion as numbered species in *Flora Europaea* might well be explained as a taxonomic "accident" arising from the fact that relatively few local species have been described in the genera concerned. A second group of micro-species, represented by *Bromus*, *Epipactis* and at least some species of *Euphrasia*, show critical taxonomy associated with habitual self-pollination, and it is tempting to see here a causal relationship (though one has to admit that there are some serious unsolved problems in postulating such micro-evolutionary speciation). These two groups account for 14 of the 23 species.

Seven of the remaining nine species belong to groups in which allopolyploidy and hybridization seem to have been involved in their recent evolutionary history. They can be considered individually, in alphabetical order. *Fumaria occidentalis* and *F. purpurea* Pugsley, both first described by Pugsley at the beginning of the present century, belong to the largely Mediterranean section *Grandiflorae* of *Fumaria*. *F. occidentalis* impressed Wilmott (1930) as about the clearest case of a British endemic species; Daker (1964) gives convincing evidence for its origin by allopolyploidy, and considers, though more hesitantly, that *F. purpurea* has had a similar evolutionary history. *Primula scotica*, described by Hooker in 1821, is certainly distinct from other British primulas, but closely resembles *P. scandinavica* Bruun, endemic to Scandinavia. Slight morphological differences, combined with a different chromosome number, have sufficed, however, to make modern authors follow Wright-Smith and Fletcher (1938) in confirming its specific rank (see also Ritchie, 1954). Both *P. scotica* and *P. scandinavica* could be considered as polyploid derivatives of the widespread European species *P. farinosa* L. *Saxifraga hartii* D. A. Webb, first distinguished by Webb in 1950, is said by the author to be intermediate in morphology between the Arctic *S. cespitosa* L. and the related *S. rosacea* Moench of European mountains; it could be "the relict of a hybrid population", confined to Arranmore Island off the Irish coast. The whole group is polyploid, with chromosome numbers ranging from $2n = 50$ to 80.

Senecio cambrensis Rosser is unique; it is a new allopolyploid species which has arisen by hybridization between the invasive ruderal *S. squalidus* L. and the widespread weed *S. vulgaris* L. Plants indistinguishable from the wild *S. cambrensis* were synthesized by artificial crossing and chromosome doubling (Rosser, 1955). Recent records suggest that this new, fertile allopolyploid is spreading from its original locality, and it seems probable that the species will, like the famous allopolyploid *Spartina anglica* C. E. Hubbard, soon cease to be a purely British endemic, and take its place in the flora of continental Europe.

Gentianella anglica (Pugsley) E. F. Warburg and *Salix hibernica* Rech. fil. are both found in a "group" in *Flora Europaea*, a treatment which indicates more than the usual degree of taxonomic difficulty in distinguishing the members. *G. anglica* was first distinguished as a British variety of the widespread *G. amarella* (L.) Börner, and raised to specific rank by Pugsley in 1936. The specific status was adopted by Pritchard (1959), who carried out a detailed, experimental study of the group and was the author of the account in *Flora Europaea*. *Salix hibernica*, known to Irish and British botanists for many years as a peculiar form of *S. phylicifolia* L. on Ben Bulben, Co. Sligo, was described as a species by Rechinger (the author of *Salix* in *Flora Europaea*) in 1963; it is clearly part of the complex, variable group of taxa related to *S. phylicifolia*, itself a high polyploid species (2*n* = 114).

These two taxa obviously constitute border-line cases, where the treatment of the endemic variant could easily have been at the subspecific or varietal levels. They form a link to several endemic British taxa which have received in *Flora Europaea* only infraspecific recognition or have been included in synonymy. Examples are *Aconitum anglicum* Stapf., the wild Monkshood of Britain (treated as conspecific with the variable *A. napellus* L. of continental Europe); *Cerastium edmondstonii* (H. C. Watson) Murb. & Ostenf., endemic to serpentine rock in the Shetland Isles (treated as a subspecies of the Arctic *C. arcticum* Lange); *Linum anglicum* Miller, endemic to calcareous grassland in Britain (treated as a subspecies of *L. perenne* L.) and *Allium babingtonii* Borrer, a famous, doubtfully native coastal species of Cornwall, the Scilly Isles and N.W. Ireland (reduced to a variety of the variable species *A. ampeloprasum* L). A unique case is *Centaurium latifolium* (Sm.) Druce, an endemic of sand-dunes in N.W. England, well known in the nineteenth century but now apparently extinct; this receives mention in *Flora Europaea*, but only as "probably a mutant" of the common and variable *Centaurium erythraea* Rafn.

Conversely, recent research has in a few cases resulted in a British population of a particular variant being raised to the status of an endemic subspecies in *Flora Europaea*. Good examples are the Breckland *Scleranthus perennis* L., now distinguished as subsp. *prostratus* P. D. Sell; the Teesdale variant of *Helianthemum canum* (L.) Baumg., now distinguished as subsp. *levigatum* M. C. F. Proctor; and the very remarkable endemic *Senecio* on seacliffs in Anglesey, now distinguished as *S. integrifolius* (L.) Clairv. subsp. *maritimus* (Syme) Chater. These cases are particularly interesting, because in each of them part of the British population referable to the European species is found to belong to a non-endemic continental subspecies, so that the pattern is not one of island endemism, but rather of specialized endemic topodemes (local populations) spatially isolated from the rest of the species to which they are referred. Such infraspecific endemism might be expected to occur frequently within continental Europe, and in fact is to be found in the case of all three of these species. It is obviously associated with eco-geographical specialization and is part of a general pattern shown by many widespread European species with partly discontinuous ranges. Its manifestation in Britain in these particular cases merely emphasizes how the British flora as a whole is to be understood as a relatively depauperate part of the European flora, and the geologically recent island status seems to play very little part indeed in micro-evolution arising from isolation.

CLASSIFICATION OF ENDEMICS: COMPARISON WITH CONTINENTAL WORK

As pointed out by Prentice (1976), most classifications of endemics involve speculation as to the evolutionary history of the taxon concerned. The largely cytologically based classification due to Favarger and Contandriopoulos (Favarger, 1972, 1974, and references given there) does, however, to a degree avoid this subjective approach, at least so far as their categories other than the "palaeoendemics" are concerned. In this classification the putatively ancient, relict endemics of very clear taxonomic status are the palaeoendemics: as we have said, the British flora contains no species which could reasonably be assigned to this group. The second group contains the so-called "schizoendemics", which are presumed to have arisen by gradual speciation and have remained at the diploid level. The two endemic *Rhynchosinapis* species, both of which have $2n = 24$, would seem to be the only cases which might be assignable to this group, although, since the widespread *R. cheiranthos* (Vill.) Dandy has $2n = 48$, the micro-evolutionary history of the genus as a whole has clearly involved polyploidy. Indeed the third group, "patroendemics", defined as restricted diploid endemics which give rise to widespread polyploids, might fit the British *Rhynchosinapis* cases better. Finally, the so-called "apoendemics" are polyploid species of restricted area which have arisen from widely distributed diploids, and it is clear that virtually all the British endemics, including the many apomictic micro-species, fall into this group. The known exceptions are: *Bromus pseudosecalinus* P. M. Smith ($2n = 14$), a recently-distinguished species which may well on further investigation be shown to occur outside the British Isles, and the Euphrasiae, *Euphrasia rivularis* Pugsley and *E. vigursii* Davey, which belong to the diploid group of species related to *E. rostkoviana*

Hayne. Yeo (1956) has discussed in particular the possible hybridogenous nature of *E. vigursii*, which may well be a special case of considerable theoretical interest. *E. rivularis*, on the other hand, a small-flowered mountain endemic, might well have originated from a *"rostkoviana"*-like ancestor; it could therefore be tentatively placed in the "schizoendemic" group.

Polyploidy and apomixis are therefore the phenomena obviously underlying most of the small amount of endemism detectable in the British flora. The patterns of micro-evolution are those associated with recently glaciated regions of northern Europe or the European Alps, and cannot reasonably be thought to have much to do with the very recent separation of the British Isles from the European continent. Indeed, as we have already seen, comparable endemic taxa can be found within the same genera to which our endemics belong in various parts of continental Europe. The resemblance is in many cases a resemblance of specialization to a particular kind of habitat (sea-cliff, serpentine rock etc.). Endemic taxa, presumed to have arisen from more widespread ones, are normally more specialized in their habitat requirements, and there is nothing surprising in this.

CONSERVATION OF ENDEMIC TAXA

We have seen that there is in the British Isles a small, but by no means negligible, endemic element in the flora. Compared with the rich endemic floras of Mediterranean mountain ranges or islands, it may seem insignificant, but, since the genetic material which it represents is obviously unique, we should be concerned to afford it special protection. Morevoer, the theoretical interest in the material is very great, for we see from the relatively crude analysis performed here that certain generalizations emerge rather clearly from a bio-systematic study of our endemic species. More study is obviously needed, both experimental investigation and detailed autecological observations, on most of the species in the Appendix, partly to further the elucidation of their evolutionary history, but also, more urgently, so that we can protect and manage the populations which are often very limited in area or numbers of individuals. This concern for the endemic element in our native flora is growing in Europe as a whole (see, for example, Walters, 1971; Favarger, 1974), and there is a great need to look upon the problem as one of international importance, though normally a national conservation responsibility. The existence of the *List of Rare, Threatened and Endemic Plants for the Countries of Europe* (Lucas and Walters, 1976) now enables us to translate our international concern into action at the European, the national and the local levels.

REFERENCES

Clapham, A. R., Tutin, T. G. and Warburg, E. F. (1962). "Flora of the British Isles", 2nd edition. Cambridge University Press, Cambridge.
Daker, M. G. (1964). Cytotaxonomic studies of European *Fumaria*. Ph.D. Thesis, University of Wales.
Favarger, C. (1972). Endemism in the montane floras of Europe. *In* "Taxonomy, Phytogeography and Evolution" (D. H. Valentine, ed.), pp. 191–204. Academic Press, London and New York.
Favarger, C. (1974). Progrès récents dans l'étude de l'endémisme végétal en Europe. *Lavori Soc. Ital. Biogeog.* n.s. 4, 5–29.
Gilbert, J. L. (1967). Britain's endemic flowering plants. *Gard. Chron.* 162, 16–18.
Godwin, H. (1975). "The History of the British Flora", 2nd edition. Cambridge University Press, Cambridge.
Good, R. d'O. (1974). "The Geography of the Flowering Plants", 4th edition. Longman, London.
Linnaeus, C. (1753). "Species Plantarum". Stockholm.
Lucas, G. Ll. and Walters, S. M. (1976). "List of Rare, Threatened and Endemic Plants for the Countries of Europe". IUCN, Kew, London.
Matthews, J. R. (1955). "Origin and Distribution of the British Flora". Hutchinson, London.
Prentice, H. C. (1976). A study in endemism: *Silene diclinis. Biol. Conserv.* 10, 15–30.
Pritchard, N. M. (1959). *Gentianella* in Britain, I. *G. amarella, G. anglica* and *G. uliginosa. Watsonia* 4, 169–193.
Ray, J. (1670). "Catalogus plantarum angliae". London.
Ritchie, J. C. (1954). *Primula scotica* Hook *J. Ecol.* 42, 623–628.
Rosser, E. (1955). A new British species of *Senecio. Watsonia* 3, 228–232.
Salisbury, E. J. (1932). The East Anglian flora. *Trans. Norf. Norw. Nat. Soc.* 13, 191–263.
Turrill, W. B. (1948). "British Plant Life". Collins, London.
Tutin, T. G., Heywood, V. H., Burges, N. A., Valentine, D. H., Walters, S. M. and Webbs, D. A., eds (1964→). "Flora Europaea" 5 vols (in progress). Cambridge University Press, Cambridge.
Wallace, A. R. (1895). "Island Life", 2nd edition. Macmillan, London.

Walters, S. M. (1971). Index to the rare endemic vascular plants of Europe. *Boissiera* **19**, 87–89.

Walters, S. M. (1972). Endemism in the genus *Alchemilla* in Europe. *In* "Taxonomy, Phytogeography and Evolution" (D. H. Valentine, ed.), pp. 301–305. Academic Press, London and New York.

Wilmott, A. J. (1930). Concerning the history of the British flora. *Soc. de Biogéographie* (*Paris*) **3**, 187–190.

Wright-Smith, W. and Fletcher, H. R. (1943). The genus *Primula*, section *Farinosae*. *Trans. R. Soc. Edin.* **61**, 1–65.

Yeo, P. F. (1956). Hybridization between diploid and tetraploid species of *Euphrasia*. *Watsonia* **3**, 253–269.

APPENDIX

List of species endemic to the British Isles

Based on Lucas and Walters (1976), and including only those species which are "numbered"—i.e. fully accepted—in *Flora Europaea* (Tutin *et al.*, 1964→). (Br) = Britain only; (Hb) = Ireland only; (Br,Hb) = Britain and Ireland.

Alchemilla minima Walters (Br)

Bromus interruptus (Hack.) Druce (Br)

B. pseudosecalinus P. M. Smith (Br,Hb)

Calamagrostis scotica (Druce) Druce (Br)

Epipactis dunensis (T. & T. A. Steph.) Godfery (Br)

Euphrasia campbelliae Pugsley (Br)

E. marshallii Pugsley (Br)

E. pseudokerneri Pugsley (Br,Hb)

E. rivularis Pugsley (Br)

E. rotundifolia Pugsley (Br)

E. vigursii Davey (Br)

Fumaria occidentalis Pugsley (Br)

F. purpurea Pugsley (Br,Hb)

Gentianella anglica (Pugsley) E. F. Warburg (Br)

Limonium recurvum Salmon (Br)

L. paradoxum Pugsley (Br,Hb)

L. transwallianum (Pugsley) Pugsley (Br,Hb)

Primula scotica Hooker (Br)

Rhynchosinapis monensis (L.) Dandy ex Clapham (Br)

R. wrightii (O. E. Schulz) Dandy ex Clapham (Br)

Salix hibernica Rech. fil. (Hb)

Saxifraga hartii D. A. Webb (Hb)

Senecio cambrensis Rosser (Br)

Chapter 14
European Floristics: Past, Present and Future

V. H. HEYWOOD

INTRODUCTION

Floristic studies range from the preparation of partial lists of the plants collected in a particular area, through complete check-lists, to detailed national or regional Floras. Such is the diversity of the European flora and the imbalance in our knowledge of its various regions and elements, that the range of floristic studies carried out today is as great as two centuries ago, although the emphasis is very clearly in the preparation of Floras.

Floristic work has a long and honourable tradition in Europe, and both amateur and professional botanists have played their role, often collaborating fruitfully either on a personal level or through the aegis of one of the numerous botanical or natural history societies that have been a feature of the scientific–cultural life of so many European countries.

The completion of the *Flora Europaea* project, of which Professor T. G. Tutin has been Chairman for the past 23 years, is an appropriate occasion on which to review the pattern of floristic work in Europe. After a brief historical survey, an attempt will be made to assess the present day situation, and consider what likely and desirable developments in the future might be.

HISTORICAL

European floristic studies can be regarded as having started in the sixteenth century when modern botany developed out of herbalism as an independent study (Stearn, 1975). Until then, and for some time subsequently, most interest and emphasis was placed on the association of botany with medicine, through *materia medica* and herbals. Attempts were made to extend the range of plants available as *materia medica* by giving vernacular names which equated with the Latin equivalents.

The number of species recorded in herbals and used in *materia medica* was initially very limited, and restricted to those plants described in traditional works and handed down from generation to generation. Few botanic gardens had been founded, and these in turn were poor in the range of plant species—primarily medicinal and culinary plants were cultivated.

The position in the sixteenth century, then, as regards floristic botany, was that the major need was for botanists to describe and record the plants of their own native country. This was a slow business, even for countries like Britain with relatively poor floras; according to various sources, it appears that, by 1600, about 500 species of British plants were known and described; by 1700, as the result of the special effort of medical doctors, apothecaries and botanists in organizing excursions, and the extensive travel and research of eminent figures such as Thomas Johnson, John Ray and their friends and colleagues, the number of species known from Britain reached nearly 1000, and the following hundred years brought the total to 1145.

In contrast, other European countries remained largely unknown floristically at the beginning of the nineteenth century, noted examples being Spain and Greece—both of them with largely Mediterranean climates, varied geology and topography, and, as we know today, with exceedingly rich floras, comprising very many endemic species. Although Greece has been regarded as the cradle of scientific botany, most of the knowledge acquired during the classic period, as represented today by Theophrastos and Dioskorides, was lost (cf. Greuter, 1975). Although the influence of the classic writers survived and indeed influenced herbalism and *materia medica* in the rest of Europe for many decades, the exploration of the Greek flora was a slow and arduous process, involving naturalists, field botanists, amateurs and professionals from all over Europe, and still remains highly incomplete to the present day (Greuter *et al.*, 1976).

Spain, at the other end of the Mediterranean, retained practically nothing from the classic period or from the *materia medica* of the Arab conquest. Spanish botanists were active during the seventeenth century, the pre-Linnaean *Flora Española* of Quer being a major work. However, the size and complexity of the flora, combined with political troubles, resulted in the greater part remaining undiscovered until the voyages of the great naturalists and botanists, such as Boissier, Barker Webb, and Willkomm and Lange who later wrote the first and, lamentably, the only, complete Flora of Spain to date (1861–1893).

A curious phenomenon in the sixteenth and the earlier part of the seventeenth century was the concentration and virtual restriction of botanical activity to an area of central and southern Europe, from Salerno in southern Italy northwards through Switzerland, France and Germany to London, eastward to Prague and westward to Paris, Lyon and Montpellier. This was related to the academic and cultural development of Europe at that time, and to the establishment of medical schools and associated physic gardens (later botanic gardens), where botany tended to flourish.

The floristic exploration of Europe was, therefore, very uneven, and it makes little sense to try and trace it on a purely national basis. The patterns of scientific change and development, added to countless political upheavals and realignments, not to mention imperial and colonial activities of European powers, have combined with other accidental factors to give floristics a highly complex interrelated history in Europe. Some of the main features of the post-Linnean period to the present day will be sketched in before considering in more detail scientific aspects such as the design of Floras and concepts of taxa.

The Linnaean period caused a certain displacement of the centre of floristic activity to Sweden and Scandinavia, largely due to the enormous labours of Linnaeus himself and of his pupils. In addition to *Species Plantarum*, which in some ways can be regarded as a European Flora because of its European bias and origins (cf. Heywood, 1974), although it purported to cover the plants of the world then known to Linnaeus, works such as *Flora Lapponica* (1737), *Flora Suecica* (1745) and various floristic dissertations published on behalf of his pupils, focused attention for some years on the Swedish master.

THE FIRST AGE OF FLORAS

In the early part of the nineteenth century, while a great deal of effort went into the preparation of works of synthesis, such as De Candolle's *Prodromus*, the floristic exploration of Europe intensified, and, by the middle of the century, we had entered into what might be called the first great age of Floras. This age of Floras was not restricted to Europe, but, on the contrary, was world-wide, due to the preparation by European botanists of Floras of colonial and imperial territories. It is important to realize that, for one-and-a-half to two centuries, European taxonomic and floristic resources and efforts have been largely devoted to extra-European territories. Not only so, but within Europe there has been a certain taint of imperialism in the way in which the exploration and description of some of the richer European floras has been carried out by botanists from other countries with scant regard for the establishment of collections and traditions. The history is, however, a mixed one, as we shall see.

British botanists, during the nineteenth century period of expansion, concentrated on the preparation of Floras of the British Empire—examples being W. J. Hooker's *Flora Boreali-Americana* (1829–1840); G. Bentham's *Flora Hongkongensis* (1861) and *Flora Australiensis* (1863–1878); and Grisebach's *Flora of the British West Indian Islands* (1850–1864), to mention just a few from a remarkable array. The botanists of other European imperial or colonial powers, such as Belgium, France and Germany, were similarly engaged.

Within Europe, floristic activity underwent an almost explosive development, especially in the countries bordering the Mediterranean. The first major explorations of Spain were made by botanists such as Boissier, whose *Voyage Botanique dans le Midi de l'Espagne* (1839–1845) was a magisterial work which transformed our knowledge of the Andalucian flora. In Greece, following the expedition of John Sibthorp, accompanied by the botanical artist Ferdinand Bauer, leading eventually to the publication of Sibthorp and Smith's *Flora Graeca* (1806–1840), and after many collecting trips by naturalists such as Sieber and Grisebach, botany was transformed by the arrival of a young German botanist, Theodor von Heldreich, who, persuaded by Boissier, stayed in Greece for nearly 60 years to become the country's most renowned botanist, collector and explorer. The first modern Flora of Greece (although excluding major parts of the modern Greek State) was written by the Austro-Hungarian botanist and physician, Eugen von Halácsy (1900–1908, 1912).

And so the pattern was set in much of Mediterranean Europe—the local effort was often eclipsed by an influx of foreign botanists, professional and semi-professional, and many amateurs. Renowned amongst these were Boissier, Bornmüller, Cosson, Gandoger, Lacaita, Maire, Lange, Reuter, Willkomm, Webb and many others. The amateurs, often highly skilled, included Bicknell, who wrote a *Flora of Bordighera* (1896), and Moggridge who is remembered for the *Flora of Mentone* (1867). In addition, several professional collectors played an important role in floristic exploration, and in making valuable material, often of new species, readily available to European herbaria and taxonomists. These collectors usually set very high standards and, with some exceptions (such as Reverchon), were highly reliable. They included Bourgeau, Balansa, Porta, Rigo, Reverchon and Sintenis.

A serious consequence of foreign predominance in the study of Mediterranean floras was pointed out by me on a previous occasion (Heywood, 1961), with reference to the *Prodromus Florae Hispanicae* of Willkomm and Lange, and deserves repeating here because it is of more general application:

The fact that the *Prodromus* was written by non-Spanish botanists and largely based on the work and collections of foreign workers had certain unfortunate consequences. Many of the basic collections, particularly types, were not easily available to Spanish botanists so that the interpretation of the species as described by Willkomm and Lange was difficult to undertake. The foreign authors did not themselves pay sufficient attention to material in Spanish herbaria. Two separate traditions grew up: on the one hand foreign botanists worked on the Spanish flora on the basis of the 'classic' collections, cited very largely in the *Prodromus*, which were available in their national herbaria; they were usually ignorant of the Spanish literature and collections; on the other hand Spanish botanists built up their own separate collections and literature and laid great emphasis on the material of Cavanilles, Lagasca and later Spanish authors. One of the great problems today is the unification of these two trends into a coherent whole, which will form the only satisfactory basis of a modern comprehensive Flora of Spain.

The taxonomic reappraisal which this will involve necessitates consideration of the enormous amount of floristic and nomenclatural studies on the Spanish (and the Portuguese and N. African) flora published since 1870, including the prolific output of Carlos Pau, and such controversial figures as Gandoger, Hervier, Reverchon, Sennen and others. The very magnitude of this task (involving the problem of type material and a bibliography of marathon proportions) has so far postponed any attempt to produce a new Flora of Spain.

In other parts of Europe, very many Floras—national, regional or local—were compiled during the nineteenth century. These run to many hundreds, often appearing in regularly revised editions, and vary so widely in scope, form, content and size, that no simple generalization can be made about them.

Many of the leading botanists of Europe were involved in this mammoth Flora-writing exercise, but these Floras seldom appear to have been regarded as major official undertakings, but rather as by-products of busy careers. Thus J. D. Hooker's *The Student's Flora of the British Islands* and G. Bentham's *Handbook of the British Flora*, later revised by Hooker, and then by Rendle and altogether surviving in its various editions nearly a century, can scarcely be regarded as major commitments for their authors when one takes into account the numerous other floristic and taxonomic enterprises on which they were also engaged. And the same is true for the leading botanists in other centres such as Vienna, Paris, Berlin etc., and it seems to be a reasonable generalization that national floristic work and Flora-writing has seldom been a primary occupation of what one might term official European botany, although many leading figures have participated in, or sponsored, such activities.

Certainly the national Floras were written, perhaps even too many and often of poor quality, and there seemed to be no shortage of students of floristics. Much of the effort fell into the hands of botanical or natural history societies, and remarkably skilled amateurs, a tradition that happily persists to the present day.

Attempts to synthesize on a European scale the great amount of floristic information being published were made by Nyman in his *Sylloge* (1854–1855) and, later, his *Conspectus* (1878–1882), the latter of which remains remarkably useful to the present day. Towards the end of the nineteenth century, a start was made on other major works of synthesis on regional Floras, the most notable being Ascherson and Graebner's *Synopsis der mitteleuropäischen Flora* (1896–1938), followed in 1906 by the first part of one of the most remarkable Floras of modern times, Hegi's *Illustrierte Flora von Mitteleuropa* (1906–1931; 2nd edition, 1936→).

The twentieth century, until the post-war 1950s, was of mixed fortune for European floristics. Several of the great Floras were published or at least initiated, such as Hayek's *Prodromus Florae Peninsulae Balcanicae* (1924–1933), Coste's *Flore de la France* (1900–1906), Cadevall's *Flora de Catalunya* (1913–1937), Halácsy's *Conspectus Florae Graecae* (1900–1904), and Komarov's 30-volume *Flora URSS*, containing 22,000 pages of text and 1250 plates. An account of the preparation of this remarkable Soviet Flora is given by Heywood and Bobrov (1965).

THE MODERN PERIOD

The combined effects of the world wars and the rise of experimental botany on the floristic scene in Europe was dramatic. Taxonomy as a whole fell from fashion; its teaching was drastically reduced or even eliminated and taxonomic higher degree dissertations were forbidden in some universities. The national institutions for taxonomy continued, largely unpublicized, their overseas Floras and monographic studies. During the 1939–1945 world war, European floristics made little progress, although a remarkable number of botanists managed to work away quietly under difficult circumstances. Perhaps the major publication was Rechinger's *Flora Aegaea* (1943).

In 1944, when it seemed that the outcome of the war was still likely to be in favour of the Axis powers, the German botanist Werner Rothmaler proposed a highly ambitious project to prepare a *Flora Europaea* embracing not only Europe but parts of North Africa. The project was stillborn, but it had its influence on the post-war generation of taxonomists, working to pick up the threads and looking for new orientations for their work. Taxonomy was largely dominated, at least in the universities, by the ideas of the new systematics, experimental taxonomy, biosystematics, and other manifestations of the application of genetics, cytology and statistics to the study of variation and classification.

A new factor which had far reaching effects on European taxonomy and floristics was the rise of the subject "ecology". In much of central and southern Europe, ecological studies were specialized in that area known as phytosociology which, in many ways, can be regarded as a derivative of floristics, and shares a common philosophy with it to the extent that it contains a classificatory component. Some countries, notably the United Kingdom, opted out of the phytosociological approach, preferring a much less structured and much more informal pragmatic approach to vegetation description and classification, while ecologists tended to devote their efforts to dynamic or experimental ecology, closely related to and largely derived from experimental taxonomy on the one hand, and whole plant physiology on the other.

It was highly unfortunate that phytosociology flourished in those countries of the Mediterranean where the flora was least well known—notably Spain, Italy, Greece, and, to some extent, France. The difficulties of undertaking phytosociological work in the absence of accurate floristic information are well known, but the major drawback for the development of floristics and taxonomy in these countries was the simple fact that young botanists were encouraged to take up phytosociology to the neglect of taxonomy, a fact which has only belatedly been recognized. Training in taxonomy and floristics virtually ceased in some countries and even the traditions were lost.

One of the problems created by the activities of the phytosociologists was the publication of apparent floristic information in their relevés, often the only source of information for many areas. In the absence of documentation, and in the context of inadequate taxonomic background, it is exceedingly difficult to know how to interpret such floristic information in relevés. In recent years the situation has improved substantially, but a large problem remains.

THE *FLORA EUROPAEA* PERIOD

In the early 1950s a number of botanists, mainly British, became concerned at the practical effects of the then current neglect of taxonomy. These were most notable in ecology and plant geography, especially when attempts were made to view relationships on a wide scale. It was apparent that far too much taxonomy had been done on a national basis, a tendency encouraged by the laborious techniques of experimental taxonomy, whereas the plants themselves recognized no boundaries. Thus the same species could be seen to be treated differently in different Floras because an essentially national or local view-point was being taken. These factors, combined with the realizations that synthesis of floristic information required a great deal of complex research, compilation and correlation, and the review of thousands of books and papers in more than a score of languages, suggested that there was a need for setting an organization to prepare an overall European synthesis of the European flora. A discussion on these and related topics was held at the International Botanical Congress in Paris in 1954, but no agreement was reached on the question of starting work on a European Flora.

Subsequently the *Flora Europaea* organization was set up as the result of the initiatives of a small group of British botanists. The first meeting of the Editorial Committee was held in January, 1955, and the final formal meeting was held in September, 1977. Details of the background to the *Flora Europaea* project and its early history have been published elsewhere (Heywood, 1957, 1958) and I shall not go over the ground again. It is too near the event to give an assessment of what *Flora Europaea* has achieved, but a great many lessons have been learned by the editors about the conduct of European floristic botany, and some of these will now be discussed.

Flora Europaea represented a new departure in the writing of Floras. The work was not only cooperative, relying on scores of authors, and a panel of regional advisers, representing each European state, who had the opportunity of commenting on all the manuscripts in the light of their specialized local knowledge, but was under very tight editorial control. This editorial control, exercised by the Editorial Committee, whose individual members each took primary responsibility for particular families, meant in practice that all manuscripts were extensively reviewed, modified and revised until they met the norms and standards which had been worked out over a period of years. This editorial process continued through all stages of processing of the account, from the first manuscript when checks were made as to completeness, general accuracy, effectiveness of keys, synonymy, distribution and style, even before the initial distribution, through the various stages of circulation to the advisers until the final checking before going to press.

While authors were consulted at all stages, the final responsibility for the published text lies with the editors. Fortunately, with very few exceptions, the authors saw the advantages of subsuming their own individual effort, and indeed pride, to the combined experience of all specialist and local advisers. The resultant text is much more accurate than it would otherwise have been, although it often involved sacrificing some of the author's idiosyncrasies. This system of editing has sometimes been criticized by reviewers who cannot accept the principle of subservience of the individual to the communal effort under editorial control. Yet on analysis it is difficult to see how else, in practice, a Flora of Europe could be written. It has to be accepted that a one-man Flora of Europe or a comparable area is no longer possible so that a multiauthor work is a necessity. From this it follows that a greater degree of editorial control is needed for the sake of consistency and uniformity. But the matter does not end there, for, as we have already seen, the wealth and diversity of European floristic resources are such that no individual author can hope to have access to them all, even assuming that the knowledge and time were available.

It has in fact taken the Editorial Committee of *Flora Europaea* 20 years to build up the vast amount of diverse skills, expertise, specialist and often esoteric knowledge about European taxonomy, taxonomists' collections, herbaria, types, nomenclature, geography, and so on, to say nothing of understanding the different national traditions and philosophies as well as the peculiarities of individual taxonomists, all of which was necessary to carry out the task of editing effectively.

Writing a European Flora is a very different exercise from that of a research Flora. In the latter case, the documentation—specimens and literature—is limited and not only personally verified by the author concerned, but also noted and recorded in the text of the published Flora. The corresponding data-base for Europe is of quite unmanageable proportions except in the case of species of restricted distribution. The herbarium specimens are distributed throughout hundreds of herbaria, large and small, mostly but by no means all in Europe. Except in the case of local taxa, the normal procedure is for the author to consult as many as possible of the major national herbaria (the choice depending on the overall distribution pattern of the genus concerned) which will normally provide a rich collection of material for study. In addition the relevant literature is scattered throughout hundreds or thousands of books or papers in scientific journals in a score or more languages. This vast literature represents the cumulative experience of two centuries of taxonomists and cannot be ignored. Yet it is exceedingly difficult to find this literature, let alone attempt to read, digest and collate it. It is for this reason that the system of review of manuscripts by national or regional advisers was instituted—so providing the author with access through local specialists to national or local literature and material. A similar approach might well be considered for the writing of monographs and revisions.

One of the most difficult problems one comes across in attempting to correlate the information given in different Floras is the variation in species-concepts. This can be so extreme in some cases as to make comparisons virtually meaningless, as I pointed out on a previous occasion (Heywood, 1967). Although something approaching a consensus between leading European taxonomists about the application of the concepts of species, subspecies and variety was reached in the 1950s, this was not, of course, retrospective in effect and left taxonomists with difficult equations such as how many of Fiori's formae are equivalent to Rouy and Foucaud's subspecies (or proles) or Coste's species?

One significant tendency in recent years has been the simplification of the infraspecific hierarchy used in Floras. While so-called critical Floras often included subspecies, varieties and forms, an increasing number of Floras now tend to recognize only one infraspecific variant, normally the subspecies. This largely reflects a realization that the nature and pattern of infraspecific variation is so complex and often multidimensional that it cannot be adequately accommodated in a rigid set of formal categories such as subspecies, varieties etc.

On the other hand, the problem of handling infraspecific variation remains largely unsolved and no effective system of referring to the various kinds of population variants—ecotypic, genetical, cytological, morphological, chemical etc. has yet been devised.

Another tendency has been the recognition of narrow species, equivalent to subspecies or varieties of more conventional treatments. This has been most marked in the U.S.S.R., notably in Komarov's *Flora URSS*, and in several East European countries, although this practice is by no means confined to these areas.

These two tendencies—the widespread use of subspecies as the only formal infraspecific variant and the acceptance of a narrow species concept (often both in the same Flora) has been referred to as the inflation of the taxonomic categories. It is, perhaps, no more than part of a cyclical pattern as is seen in the comparable current trend towards the recognition of small genera, recognizable by micro-morphological or technical characters, segregated from the larger genera that were fashionable previously, although analysis has shown that the small genera had often been recognized in even earlier periods.

Apart from the above mentioned changes in the usage of the taxonomic categories the basic content of Floras in the post-war period to date has shown little change. The most commonly given additional information is chromosome numbers. In most Floras published recently the chromosome number information is very limited, usually without documentation of the counts given and in many cases there is not even an indication if the numbers refer to native material collected within the area covered by the Flora.

Ecological and geographical information is again somewhat restricted, except in research Floras where specimens are cited. The distribution of taxa within the area of the Flora is sometimes given in considerable detail, occasionally using some form of grid reference and sometimes with distribution maps. Overall distribution of taxa may be indicated in general terms but often this information is copied from other works and is not fully reliable.

Several recent European Floras include phytosociological data about the taxa, such as Oberdorfer's *Pflanzensoziologische Excursionsflora* (1970), Rothmaler's *Excursionsflora* (1972), Soó's *A Magyar Flóra és Vegetáció Rendszertani-növényfoldrajzi Kézikönyve (1964–1973)* and Guinochet and Vilmorin's *Flore de France* (1973→). These Floras are able, by reference to established phytosociological classifications, to convey a large amount of detailed information in abbreviated form.

As regards morphological information, little advance has been made in the presentation or quality of information, although more attention has been paid to the style, sequence and comparability of descriptions and to the effectiveness of keys. The guide produced for contributors to *Flora Europaea* entitled *The presentation of taxonomic information* (Heywood, 1958) has played a significant role in this respect and has been widely used by Flora writers both in Europe and in other parts of the world.

Despite the widespread availability of micro-morphological data, especially derived from scanning electron microscopy, referring to seeds, spores, pollen grains and indumentum, this information is rarely incorporated into Floras. This is an area that will require considerable attention in the design of Floras for the future. One of the few Floras which has attempted to include information and references to these more recent kinds of data, including pollen structure and chemical constituents, is the second edition of Hegi's *Illustrierte Flora von Mitteleuropa*. The problems of presenting all this information in a Flora like that of Hegi have been discussed by me previously (Heywood, 1971).

THE FUTURE PROSPECT

The successful completion of *Flora Europaea*, providing an overall floristic–taxonomic synthesis and perspective, together with the considerable number of national and regional Floras published during the last two decades, has transformed the European scene. *Flora Europaea* has identified many problem groups which require further detailed study—the percentage of unnumbered species referred to in observations after the numbered species is about 10% of the total, a remarkably high percentage. Many of these "species" will no doubt be shown to be little more than variants of accepted species but it will clearly take many years for an adequate assessment of them to be made. In addition there are many critical genera where the treatment in *Flora Europaea* and in other recent Floras is provisional and in need of thorough revision.

No revision of *Flora Europaea* as a whole is at present envisaged, although there is a clear need for the establishment of some updating system, possibly using modern electronic data-processing equipment. It should be remembered that *Flora Europaea* was planned and largely written before the application of electronic data-processing techniques to taxonomic and bibliographic systems was sufficiently advanced and matured. The experience of the ill-fated Flora of North America project, which was designed as an information system using electronic data-processing has, in retrospect, proved of inestimable value and the lessons learned will make the task of future projects of this nature more viable. It is beyond doubt that some of the available machinery and systems would greatly simplify many of the processes involved in Flora-writing.

The whole question of the design of Floras requires considerable attention. Little advance has been made in practice during the last century from the time when the data available were much more restricted in kind than today (cf. Heywood, 1973a, b). Not only are procedural innovations necessary [as, too, in the preparation of monographs or revisions, cf. Meacham and Stuessy (1976)] but so are changes in the form of presentation of the various classes of data to be included in Floras.

Another area, already referred to above, where a new look is needed is the study and presentation of data on infraspecific variation. The problems involved have been recently reviewed by Stace (1976). Whether effort should be devoted to such questions in Europe is debatable when one considers the enormity of the problem facing us in tropical floras where both known and unknown plant resources are being destroyed at an alarming rate. I suspect that priority will in future be given to the Mediterranean parts of Europe and to groups of agronomic importance, as regards both floristic and variational studies. The great age of European floristics is over. It would be indulgence, I believe, to invest further substantial resources in this direction while the floras of the Mediterranean, of islands, and of the tropics, which are vastly more diverse and rich, and of enormous potential value to mankind, are subject daily to increasing threats of extinction.

REFERENCES

Greuter, W. (1975). Floristic Studies in Greece. In "European Floristic and Taxonomic Studies" (S. M. Walters, ed.), pp. 18–37, Bot. Soc. Brit. Is. E. W. Classey, Faringdon.

Greuter, W., Phitos, D. and Runemark, H. (1976). Greece and the Greek Islands. A report on the available floristic information and on current floristic and phytotaxonomic research. In "La Flore du Bassin Méditerranéen: Essai de Systématique Synthétique," pp. 67–89. Colloques Internationaux du C. N. R. S., No. 235.

Heywood, V. H. (1957). A proposed Flora of Europe. *Taxon* **6**, 33–42.

Heywood, V. H. (1958). Flora Europaea—a progress report. *Taxon* **7**, 73–79.

Heywood, V. H. (1961). "Catalogus Plantarum Vascularium Hispaniae" 1. Instituto Botánico, A. J. Cavanilles, Madrid.

Heywood, V. H. (1967). Variation in species concepts. *Bull. Jard. Bot. Nat. Belg.* **37**, 31–36.

Heywood, V. H. (1971). The new Hegi Compositae. *Taxon* **19**, 937–938.

Heywood, V. H. (1973a). Ecological data in practical taxonomy. *In* "Taxonomy and Ecology" (V. H. Heywood, ed.), pp. 329–347. Academic Press, London and New York.

Heywood, V. H. (1973b). Taxonomy in crisis? *Acta Bot. Acad. Sci. Hung.* **19**, 139–146.

Heywood, V. H. (1974). Systematics—the stone of Sisyphus. *Biol. J. Linn. Soc.* **6**, 169–178.

Heywood, V. H. and Bobrov, E. G. (1965). Preparation of *Flora URSS. Nature, Lond.* **205**, 1046–1049.

Meacham, C. A. and Stuessy, T. F. (1976). Procedural innovations in revisionary studies: computer-assisted citation of representative specimens. *Madroño* **23**, 266–273.

Stace, C. A. (1976). The study of infraspecific variation. *Current Adv. Plant Science.* **23**, 513–523.

Stearn, W. T. (1975). History of the British contribution to the study of the European flora. *In* "European Floristic and taxonomic Studies" (S. M. Walters, ed.), pp. 1–17, Bot. Soc. Brit. Is. E. W. Classey, Faringdon.

Index

Numbers with an asterisk indicate the pages on which illustrations occur.